THE AUTOMATED BATTLEFIELD

Frank Barnaby

OXFORD UNIVERSITY PRESS
1987

Oxford University Press, Walton Street, Oxford OX2 6DP
Oxford New York Toronto
Delhi Bombay Calcutta Madras Karachi
Petaling Jaya Singapore Hong Kong Tokyo
Nairobi Dar es Salaam Cape Town
Melbourne Auckland
and associated companies in
Beirut Berlin Ibadan Nicosia

Oxford is a trade mark of Oxford University Press

First published 1986 by Sidgwick & Jackson Limited
First issued as an Oxford University Press paperback 1987

British Library Cataloguing in Publication Data
Barnaby, Frank
The automated battlefield.
1. Military art and science
2. Technological innovations
I. Title
355'.02 U104
ISBN 0-19-285189-6

Printed in Great Britain by
Richard Clay Limited
Bungay, Suffolk

THE AUTOMATED BATTLEFIELD

To Wendy, Sophie and Ben

Contents

Chapter 1
The Battlefield of the Future

'On the battlefield of the future enemy forces will be located, tracked and targeted almost instantaneously through the use of data-links, computer-assisted intelligence evaluation and automated fire control. With first-round kill probabilities approaching certainty, and with surveillance devices that can continuously track the enemy, the need for large forces to fix the opposition physically will be less important.'

This quotation is from a speech made on 14 October 1969 by General William C. Westmoreland, then the US Army's Chief of Staff, to the Association of the United States Army. The General continued: 'I see battlefields that are under 24-hour real or near-real time surveillance of all types. I see battlefields on which we can destroy anything we locate through instant communications and almost instantaneous application of highly lethal firepower.'

General Westmoreland's vision is fast becoming reality because of developments in microelectronics. Microelectronics in general and computers in particular have revolutionized our lives. But these developments have revolutionized military activities to an even greater extent. In fact, new inventions in microelectronics are usually first made by military scientists and first applied to weapon systems.

Microelectronics are changing the characteristics of weapons – particularly tanks, combat aircraft, missiles and warships – beyond recognition, even though the new weapons may look rather similar to those they replace. These changes are brought about mainly because of developments in microelectronics but also because of developments in many, actually hundreds of, other technologies. Particular mention can be made of more efficient engines; more effective fuels; reductions in the size and weight of weapons for a given destructive power; smaller weapon delivery systems; the development of a whole range of new materials, from laminated armour to optical fibres; and greater adaptability of weapon-firing platforms.

Of all the advances in military technology, those with the most far-reaching consequences are developments in surveillance, in the guidance of weapons and in warhead design. Surveillance technologies are used to locate, identify and keep track of enemy forces. Guidance technologies are used to make sure that weapons fired at enemy forces are accurately guided to their targets. Warhead technologies are important to make sure that, once hit, the target – tank, aircraft, warship or soldier – is damaged sufficiently to keep it permanently out of the battle. Together with the use of computers to assist military decision-making, advances in surveillance, guidance and warhead design are transforming warfare. The purpose of this book is to show just how much, and in what ways, conventional warfare is being transformed.

We will see that, given new and foreseeable military technological advances, all forms of warfare – conventional and nuclear, tactical and strategic – between advanced countries will become increasingly automated. There is no technological reason why warfare should not eventually become completely automated, fought with machines and computerized missiles with no direct human intervention.

We can imagine an automated conventional battle taking place in an area of territory from which all people have been evacuated. An attacker invades the area with robot-driven vehicles. The other side defends it with automatic missiles. The battle takes place with no direct human involvement. The soldiers used are robots. If there is human involvement, it is remote. Humans keep well away from the battlefield. They will do so because the battlefield has become too lethal for humans.

An important military technological advance is the development of conventional munitions that are very destructive. An example of a very destructive conventional weapon now being deployed is the multiple-launch rocket system, called MLRS. Essentially a new type of artillery, MLRS fires a salvo of rockets; typically twelve rockets are fired very rapidly. Each rocket carries one of a number of types of munition. In one version of the weapon, the warhead carries a number of grenade-type munitions. When one explodes it produces a large number of metal fragments, each capable of causing a serious wound if it hits a person. A twelve-rocket MLRS salvo carries about 10,000 of these grenades. When they explode, the lethal metal fragments, travelling at high speeds, are scattered over an area equal to that of six football fields. Any human being in range is more than likely to be killed or seriously injured. Conventional weapons like these are as destructive, at least in their short-term effects, as small (low-yield) nuclear weapons.

Military experts refer to very destructive weapons as weapons of high fire-power. Because of the development of new guidance systems for missiles, these weapons can be delivered over long ranges with pinpoint accuracy.

Missiles are not only becoming more accurate; they are becoming 'autonomous'. Once fired, an autonomous missile can seek out its target, identify it and attack it without any further instructions from the person or platform that fired the missile. For example, if the pilot of an aircraft fires an autonomous missile it will seek out and attack its target without further instructions from the pilot, even though the target is a long way away. This means that once the pilot has fired the missile he can turn his aircraft round and go home. Autonomous missiles are also called 'fire-and-forget' or 'launch-and-leave' missiles. The availability of autonomous missiles that can carry warheads of great fire-power and deliver them accurately at long ranges has far-reaching and dramatic consequences.

Automated weapons in the Third World

Weapons of increasing fire-power and the missiles to deliver them accurately are being acquired, usually through the global arms trade, by Third World countries. We must, therefore, expect wars in the Third World to become increasingly violent and destructive.

The more violent wars in Third World regions become, the more likely such a war is to escalate to a nuclear war involving the superpowers. Most observers of international affairs believe that the most likely way in which a nuclear world war will come about is through the escalation of a war that begins in the Third World. This, the experts say, is much more likely than a nuclear world war starting by one superpower attacking the other out of the blue.

The most feared scenario is that a future war in, say, the Middle East will begin as a conventional war. It will then escalate to a local nuclear war in which nuclear weapons produced by the local powers are used. The war will then spread to Europe, begin there as conventional war between NATO and the Warsaw Pact, and rapidly escalate to a tactical nuclear war, involving the use of nuclear weapons in European countries. Finally, the war will escalate to a strategic nuclear war in which the USA and the USSR bombard each other with intercontinental ballistic missiles carrying nuclear warheads.

The scenario becomes more likely if the original conventional war is very violent and fast-moving. The survival of one or both combatants may then be so threatened that one side resorts to the use of nuclear weapons. And the more countries there are that have nuclear weapons, the more likely it is that a future war will involve their use. The proliferation of conventional weapons of high fire-power and accurately-guided missiles to go with them, and the proliferation of nuclear weapons are, therefore, subjects of major concern.

But the problem does not stop there. We must expect that sub-national groups will also acquire conventional weapons of increasing fire-power, the missiles to deliver them, and even nuclear weapons. Terrorism, for example, will become an ever greater menace for both developed and developing countries.

Is automated warfare credible? Offence versus defence

Automated weaponry will spread to many countries. But the

first to acquire a range of automated military systems will be the industrialized countries. Fully automated conventional warfare will first be possible in Europe between NATO and the Warsaw Pact. But will warfare between robots be taken seriously? People may argue that if war is to be waged by computers, why bother to have weapons at all? Why not simply play a computer game? If computerization reduces conventional tactical warfare in Europe to absurdity, this phase of warfare may be eliminated. The danger then is that a conflict in Europe will rapidly escalate to a strategic nuclear war.

The dangers arising from fully automated warfare are in the medium term, to be faced by our children. Before then, though, the deployment of increasingly automated weapons systems will still lead to far-reaching changes in military doctrine, particularly for European countries.

Two different ways in which military doctrines should be changed have been suggested. The first would change military doctrines in the direction of using new military technologies to attack enemy forces, particularly reinforcements being brought up to the front line, at long distances. Typically, it is suggested that enemy forces should be attacked with ballistic missiles armed with conventional warheads at distances of up to 300 or 400 kilometres behind the lines. Using current weapons, NATO's capability of attacking targets in enemy territory is limited to the use of aircraft at distances of up to about fifty kilometres. To change NATO's doctrine to enable it to attack targets at 300 or 400 kilometres would be an offensive change, or at least might be seen as such by the other side.

The second suggested way in which military doctrines should be changed using the new military technologies is to use them for purely defensive purposes to set up a defensive conventional deterrence. The suggestion arises from the interesting fact that the new technologies make defence much more cost-effective than offence. This is because, using the new technologies for detecting, identifying and tracking enemy forces, autonomous missiles, and new warheads, it is much cheaper to destroy the weapons of invasion – heavy tanks, long-range missiles, long-range military aircraft and large warships – than to buy them.

The invasion and occupation of a country can, therefore, be

made prohibitively expensive. It is for this reason that the new technologies can be used to provide an effective deterrent, based on conventional weapons, against an attack.

In summary, increasingly automated weapons will become available to many countries and to sub-national groups. The consequences of this spread depend on the time scale. If technological developments hold sway, the fully automated battlefield will be with us, at least in the industrialized countries, by about the year 2010.

Before then, automated systems will be increasingly introduced. Countries will then have to modify their military doctrines to accommodate them. The choice will be either to go for long-range offensive weapons, and their supporting technologies, or to depend on defensive weapons and adopt a doctrine of defensive conventional deterrence.

Terrorist groups will acquire automated conventional weapons of high fire-power and almost certainly, in due course, nuclear weapons. The levels of terrorist violence will, therefore, increase. Surface-to-air missiles, for example, will allow terrorists to threaten civilian aircraft. Terrorists will become increasingly able to attack and destroy targets, including large parts of cities, in industrialized countries.

The importance of low-level violence, of the terrorist or urban-guerrilla variety, will increase as warfare becomes more automated. The fully automated battle between industrialized countries, fought by robots with no direct human involvement, may seem too ridiculous to contemplate. A conflict in Europe may then rapidly escalate to an all-out strategic nuclear war.

We have so far been talking about the automation of tactical war, the main topic of this book. But, in the longer term, strategic war between the USA and the USSR may become automated. In a strategic war between the superpowers they will directly attack each other's homeland, mainly with long-range missiles. A tactical war between East and West would take place in Europe, probably beginning with one side attacking the other across the East–West German border.

The availability of very accurate missiles with intercontinental ranges equipped with conventional warheads of high fire-power will give the USA and the USSR the capability of attacking and destroying military targets in each other's territory with conventional rather than nuclear weapons. The prospect of strategic non-nuclear war has dramatic implications for East–West relations. These implications have yet to be fully explored.

A General's vision

General Westmoreland based his vision of the automated battlefield on his experiences of new military technologies used during the Vietnam War. It has become common practice for the military scientists of the advanced countries to use wars in the Third World to test new weapons and weapon systems under operational conditions. The Vietnam War was no exception. It became a laboratory for testing, among other things, a variety of sensors and missiles suitable for use in automated warfare.

Sensors, sensitive to the movements of people and vehicles, were dropped by aircraft along the probable routes of advance of Viet Cong troops. Signals transmitted from the sensors were monitored by receivers in aircraft and on the ground. Information from the sensors was used, with information from airborne and ground-based radars, infra-red and laser devices, and night-vision equipment to detect, identify, and track enemy forces.

According to General Westmoreland, the military had, as far back as 1969, 'hundreds' of different types of intelligence sensors in operation or under development. The problem, the General explained, was to 'incorporate all these devices into an integrated land combat system'. He predicted that 'no more than ten years should separate us from the automated battlefield'.

Although much has been done since 1969 to integrate automated devices into warfare, military scientists have worked less rapidly than General Westmoreland hoped. Nevertheless, the General's vision of the automated battlefield may become reality within twenty years or so. Because of automation, the battlefield will increasingly use organizations and techniques which are radically different from those used today. Military technology is, without doubt, transforming all types of warfare – conventional, nuclear, biological, and electronic.

Advances in military technology are occurring so rapidly because of the large number of scientists working in military science. Worldwide, more than 500,000 research scientists and engineers are working only on military research and development, designing and developing new weapons and their supporting technologies, and improving existing ones. This means that about 20 per cent of the world's best scientists and engineers are working for the military. If only physicists and engineering scientists, those at the forefront of technological innovation, are counted, the percentage is even higher; according to some experts, as many as a half of these scientists are working entirely for the military, producing nothing for the civilian economy.

And military scientists are given large sums of money. Worldwide, military research and development gets almost $100,000 million a year, much more than governments give for civilian research. The USA and the USSR are the biggest spenders on military science, accounting for about 85 per cent of the world total. But France, the UK, and West Germany are also significant spenders, accounting for the bulk of the remainder.

The lethality of modern conventional warfare

One dramatic transformation of warfare is the increased lethality of conventional (non-nuclear) weapons. In the words of General John A. Wickham, US Army Chief of Staff, in testimony to the US House of Representatives Armed Services Committee, on 6 February 1985, 'We are on the threshold of some enormous technological changes in conventional weapons. The yield of the weapons, the lethality of the weapons, the accuracy of the weapons means that we can build in the next four to five years, conventional weapons which will approximate nuclear weapons in lethality, and we are moving in that direction.'

Because of the increased lethality of weapons, people are killed on the modern battlefield at a rate unimaginable in past wars. And the wounded are much more seriously wounded; a typical individual casualty receives simultaneously several serious wounds of different types. If he is to survive, he needs rapid treatment from a number of doctors, each with a different speciality. Given the large number of casualties produced on the modern battlefield, the limited medical facilities available on, or near, the battlefield are soon saturated. Relatively few casualties can expect to survive. The high lethality of modern war is brought home by the words of a recent US Army report that calls for the use of 'corpse disintegrators' to cope with the disposal of the many corpses on the battlefield.

Physical wounds, serious though they will be, may not be the main cause of casualties in future wars. Psychological stress may well be a much greater hazard. Past battles were relatively slow affairs. They normally stopped at night and when the weather became very bad. And fighting typically stopped when the battlefield became covered with smoke or dust. Usually, therefore, there were many opportunities for troops to eat and rest. Future battles, however, will go on day and night, and in virtually all weathers. Dense smoke and dust will no longer limit fighting. Although future battles will generally last for a shorter time than past ones, there will be virtually no respite for the troops. Faster military vehicles will make battles much more mobile. Troops will be constantly under attack from weapons of great fire-power. The battlefield will become a nightmare of fast and unrelieved action, of deafening noise, blinding light, and mortal danger. Troops will at all times wear heavy protective clothing in case they are suddenly attacked with chemical weapons. This protective clothing will be cumbersome and hot, making fast movement difficult. The troops will become even more vulnerable to enemy fire. If people are to exist on the battlefield of the future, human engineering to immunize them against stress may be required, just as we now, as a matter of course, immunize against disease.

The lethal effects of modern weapons are extending over

increasingly large areas. People far from the immediate fighting are at risk. Consequently, in today's wars more civilians are killed than soldiers; you stand a better chance of surviving a war if you are in uniform than you do out of uniform.

Where will these trends end? Will, it is being asked, the battlefield become 'such a technologically hostile environment that soldiers themselves will not be accommodated'? There is obviously a limit to the casualty rate beyond which troops are unwilling to go into battle; in general, troops will fight only if there is a reasonable chance of survival. As standards of living improve and populations become more dependent on technology and the comforts of life, this limit increases. The lethality and destructiveness of even conventional warfare is such that the casualty rates in warfare in, for example, Europe may already be too high to be acceptable. If they are not, they soon will be. The plain fact is that humans, at least in the developed countries, are, or are becoming, too fragile to participate directly in modern warfare.

The increasing lethality and destructiveness of warfare is one reason why attention is focusing on automated warfare. As the battlefield becomes an unacceptably dangerous place, it is being said, we must minimize the number of people on or near it. Hence the growing interest in the military use of robots ('tin soldiers') instead of human troops, and of automated weapon systems.

Interest in automated warfare is also increasing because the technologies for it are becoming available. With current and foreseeable developments, there are no technological reasons why the battlefield should not be completely automated, needing no humans on it. These developments include:

new surveillance and target-acquisition systems;
increasingly intelligent missiles;
the development of very destructive conventional warheads; and
the automation of command, control, and communications operations.

Surveillance and target acquisition

The first stage of a battle is the location and identification of enemy forces. The goal is to be able to track attacking forces at as long a range as possible and in real time – in other words, to be able to follow the movements of enemy forces as they are actually happening. The enemy forces may include tanks and other armoured vehicles, aircraft, warships, military command centres, and so on.

Today's search and surveillance operations depend on the use of sensors. The sensors are normally carried on manned and remotely piloted aircraft, and satellites; they may also be buried in the ground.

Sensors may be 'active' or 'passive'. An active sensor, such as radar or sonor, transmits a beam of energy which is then reflected off the target – a tank, aircraft, warship, etc. The reflected beam is then detected by the sensor and indicates the presence and position of the target. There are also semi-active sensors that also detect radiation reflected off the target, though the radiation is transmitted, not by the sensor itself, but by the vehicle that launches the missile. A passive sensor does not itself transmit energy but responds to energy emitted by the target. This energy may be, for example, infra-red radiation emitted by a hot part of the target, such as the engine. Another type of passive sensor picks up sound waves – for example, the sound waves emitted under water by ships and submarines.

Ground-based sensors for use on the battlefield may be sensitive to, among other things, light, sound, magnetic fields, pressure, and infra-red radiation. The most common types of battlefield sensors pick up seismic disturbances in the ground caused by the movements of people and vehicles. The sensors can be implanted on the battlefield by hand, but they would most often be fired onto the battlefield by artillery or dropped by aircraft. The sensors are usually buried in the ground with just a short aerial visible. A typical seismic sensor can detect people at distances of up to fifty metres and vehicles at ten or so times this distance. Information about enemy forces collected by the sensors can be transmitted by radio over long distances. The devices can remain active for months.

The performance of a single type of sensor is limited. A seismic sensor, for example, cannot normally tell the difference between a light vehicle at a close distance and a heavier vehicle further away. But if seismic sensors are used together with acoustic sensors, which detect sounds, a more certain identification of enemy forces can be made. In fact, it would be normal to use several different types of sensor in combination. For the detection and identification of military vehicles, for example, a powerful combination would be seismic, acoustic, and chemical sensors. The latter would analyse the exhaust fumes from passing vehicles and distinguish between petrol-driven vehicles and diesel-driven vehicles. The information would be passed to computers for analysis with a good chance of identifying a high proportion of the enemy vehicles.

In the microelectronic age, very small sensors are available, even though the power supply for their transmitters is still strong enough to allow them to send signals over distances of several kilometres. Ground relay stations can be used to transmit sensor signals over much larger distances to central computers. Alternatively, relay equipment can be carried in high-flying aircraft, which may be remotely controlled, or in satellites.

By the judicious use of ground-based sensors an effective electronic border can be provided around an area of territory or between countries. The Israelis, for example, have established such an electronic border between Israel and the Lebanon capable of detecting, with high reliability, the passage of individual persons. Britain uses ground sensors for surveillance along the border between Northern Ireland and Eire. Ground sensors were used by the Americans during the Vietnam War to detect movements of Viet Cong troops down the Ho Chi Minh Trail and in a barrier around Khe Sanh.

An example of a current ground-based sensor is the British system having the rather complicated name of Terrestrial Oscillation Battlefield Intruder Alarm System, or TOBIAS. Using only seismic sensors, TOBIAS can detect a person at fifty metres (under certain circumstances the range can increase to 300 metres) and the sensors can be installed several kilometres away from the monitor. Each monitor can be used with twenty sensors simultaneously.

A more sophisticated sensor is the Remotely Monitored Battlefield Sensor System (REMBASS) that uses seismic, infra-red, and acoustic sensors, and metal detectors. The system has relay units for long-range transmission of data.

The probable presence of enemy forces can be shown at long range, while in their own territory, by surveillance sensors carried in aircraft or satellites. Short-range sensors – including electro-optical sensors (i.e. TV cameras) – can then be used to confirm the presence of the enemy forces and to identify and track them. Final target recognition may involve collecting and analysing data on, for example, characteristic infra-red emissions of various frequencies from tanks, or the characteristic sounds from ships or submarines. The use of multiple sensors – some operating on different infra-red wavelengths, others using TV, etc. – enables real targets to be distinguished from decoys deployed by the enemy to confuse the sensors.

Command, control, and communications (C3)

When enemy forces have been located and definitely identified, decisions must be made about how to deal with them. Appropriate weapons are chosen and fired at the enemy forces. This decision-making process, and the communication of the decisions to the humans or computers controlling the weapons, is called command, control, and communications, or, in military language, C3. The success of a modern battle depends on efficient C3 systems.

The information collected by surveillance and target-acquisition sensors is transmitted through the military communications network to central computers for analysis. Increasingly, these computers decide on the action to be taken against the enemy forces. On a fully automated battlefield the computers would themselves select appropriate weapons and direct them to their targets without any human involvement.

In a typical battle, the C3 operations have another task to

complete. The damage done to the enemy forces by the initial barrage must be assessed by reconnaissance using, for example, sensors carried in remotely piloted vehicles and, perhaps, reconnaissance satellites. In an automated battle, the information on the state of the enemy forces would be fed back to computers which would decide whether or not it was necessary to fire more weapons and, if so, would fire them and direct them to their targets. This assessment and re-bombardment sequence would be repeated until the enemy forces were put out of action.

In modern warfare the multitude of sensors used for surveillance, target acquisition, and reconnaissance produces such a vast amount of information that the human brain is incapable of analysing it quickly enough. It is essential, therefore, that the raw data is analysed by computer and useful information separated from useless information.

Currently, military officers receive the information passed on by the computers and participate in making decisions about which weapons to fire. Army commanders communicate their decisions to field commanders who implement them, firing and controlling appropriate weapons. But as computers become more able to make decisions, we must expect the military to use them to their full capacity. Eventually, it will be possible to computerize fully all C3 operations. Computerized C3, together with autonomous weapons, are the essential elements of automated warfare.

The weapons on the automatic battlefield

General Westmoreland talked in his 1969 speech, quoted above, of weapons 'with first-round kill probabilities approaching certainty', weapons of almost perfect accuracy. The improvement in the accuracy of weapons is one of the most dramatic consequences of military technology over the past three decades.

Norman R. Augustine, an American weapons expert, writing in 1982, calculated that the chance of a soldier hitting an enemy with rifle fire in the American Civil War was about 1 in

100,000. The odds had not increased that much up to the Second World War. The chances of hitting a target by, for example, aerial bombing in that war were no better than twenty to one, unless the target was huge. But the situation changed dramatically towards the end of the 1960s. Just how dramatically is shown by the attempts to bomb the now famous Thanh Hoa Bridge in North Vietnam. 'Over many months,' Augustine explains, '873 sorties were flown against this bridge, and 2,000 tons of conventional bombs were dropped. Still the bridge stood. Then laser-guided smart bombs were introduced. Eight sorties dropped the bridge on the first mission.'

The 'smart' bomb referred to here is a conventional bomb equipped with a sensor, that responds to laser light, and stabilizer fins, the surfaces of which can be moved to guide the bomb to its target. The target is illuminated by a laser beam. The light beam is reflected by the target and some of this reflected light is detected by the sensor in the bomb. The sensor-system passes information about the position of the target to the movable fins which guide the bomb accurately to the target.

An early laser-designated 'smart' bomb was not, by today's standards, very intelligent. In fact, its IQ was significantly less than that of a chimpanzee. The bomb could recognize its target only if it were first illuminated with a laser beam by a human operator. And the bomb had almost no memory, needing constant reminders about where the target was. But since the first 'smart' bombs, the IQ of missiles has been steadily improved. We can now foresee the 'brilliant' missile which can itself search for and recognize its target, and attack it without instructions from any external source.

Brilliant missiles will use very advanced sensors and artificial intelligence. But their development will depend most of all on microcomputers to analyse more or less instantaneously the huge amount of data they will collect. Of all the recent military technological advances, the use of microelectronics is by far the most important in the development of increasingly intelligent missiles.

Some idea of how recent technological developments have been incorporated into weapon systems can be had from modern cruise missiles. A modern cruise missile, which flies like an aeroplane, is a formidable weapon, capable of carrying a nuclear or conventional warhead over ranges of several thousand kilometres and delivering the warhead to within thirty or forty metres of its target. The missile is surprisingly small. The current American version is only about 6 metres long, about 63 centimetres in diameter, and with a wing span of about 3.6 metres; it weighs about 1,200 kilograms. Yet it is a very effective missile. The ratio of the payload (the weight of the warhead the missle carries) to the physical weight of the missile is about 15 per cent. This ratio for a ballistic missile (that flies along a ballistic trajectory, like a stone) is typically only a fraction of one per cent. In other words, the small cruise missile can carry a warhead as powerful as that carried by a ballistic missile more than fifteen times bigger.

The development of the modern cruise missile was made possible by the miniaturization of computers in terms of their volume and weight for a given capacity of the data they can absorb and process. We now know that computers need consume very little power, so that they can be made extremely small; the eventual limit to their smallness is essentially the dimensions of the molecule. Also important for the development of modern cruise missiles is the availability of accurate information about the co-ordinates of potential targets. This information has come mainly from reconnaissance satellites.

Very small but very accurate guidance systems can thus be developed. An example is the McDonnell Douglas Terrain Contour Matching (abbreviated to TERCOM system. Although the device weighs only 37 kilograms it can guide the cruise missile to within 40 metres of its target after flying a distance of 3,000 or more kilometres.

TERCOM uses a computer to compare the terrain below the missile with a pre-programmed flight plan. Details of the flight path of the missile from its launch point to its target are fed into the missile's computer before its launch, and the area under the missile is scanned by a radar altimeter while the missile is in flight. Deviations from the planned flight are corrected automatically.

The very accurate information needed to describe the position of the target and the contours of the flight path for the cruise missile's computer is obtained from maps produced by data from reconnaissance satellites. Until satellite data became available, targets could not have been located accurately enough from earlier maps to make effective use of missile-guidance systems like TERCOM, even if such systems had existed.

Although the availability of microcomputers is essential for smart weapons, other technological developments are important. Taking the cruise missile again as an example, the development of relatively small jet engines and high-efficiency jet fuels was crucial. The Williams Research Corporation, for example, has produced a turbo-fan engine that weighs a mere 60 kilograms, yet it generates a thrust of some 275 kilograms. This is remarkable for an engine that fits into a space 80 centimetres long and 31 centimetres wide.

Using their TERCOM guidance systems, cruise missiles fly very low, hugging the ground at about 20 metres over flat land, 50 metres over hilly land, and 100 metres over mountains. The missile flies so low, in fact, that it is extremely difficult to detect by radar quickly enough to fire a surface-to-air missile at it. This is why air-defence systems find it difficult to shoot cruise missiles down. The very small radar cross-section of the missile increases the difficulty. By the time enemy radars have spotted the missile, plotted its path accurately, and fired a surface-to-air missile at it, the cruise missile has probably passed out of sight.

Cruise missiles are difficult to shoot down using ground-based air-defences even though they travel relatively slowly, at subsonic speeds of about 500 kilometres per hour. Improved versions are, however, under development. These will fly at supersonic speeds.

Future navigation systems

The military importance of accurate navigation, particularly for automated warfare, cannot be over-estimated. Future missiles, and automated warfare weapons and weapon systems in general, will benefit enormously from new space-based navigational systems, like the American NAVSTAR system.

NAVSTAR is a global positioning system based on eighteen satellites orbiting in space; all the satellites should be in orbit and the system operational by 1988. NAVSTAR transmits a stream of navigational information that enables any vehicle, on land, sea, or in the air, or any missile, to determine its position and velocity with unprecedented accuracy. Using a NAVSTAR receiver, it will be possible to determine position, anywhere in the world, to within a few tens of metres in each of the three dimensions, speed to within a few centimetres per second, and time to a millionth of a second.

NAVSTAR will be used to help synchronize activities on the automated battlefield. Mating NAVSTAR with missile guidance systems will enable missiles to be guided to their targets on the automated battlefield with great accuracy. Equally important, military vehicles will be fitted with NAVSTAR receivers to assist robot drivers.

The move to unmanned aircraft

The availability of new materials, particularly new alloys, has revolutionized the performance of combat aircraft. And the most advanced electronic equipment is to be found in, for example, the air-superiority fighter, an aircraft designed to seek, find, and shoot down all types of enemy aircraft, in all weathers.

Rapid-response sensors are being increasingly used to control high-performance aircraft and to fire and control their weapons. The faster the aircraft travel the more necessary automated control becomes. And military aircraft are becoming faster all the time. The United States is considering a quick-reaction interceptor aircraft, capable of speeds greater than three times that of sound and extremely rapid acceleration. Even faster and more manoeuvrable aircraft are being discussed.

To fly and at the same time to fire and control the weapons of a modern combat aircraft is so difficult as to be beyond the capability of one man. Consequently, research is under way to produce voice-activated computer systems, so that a pilot can use facilities other than his arms and legs, which are, in any case, already fully occupied. When voice-activated systems have been satisfactorily developed, the pilot will be able to operate controls and fire weapons by talking to the aircraft's computer.

The electronic equipment in a modern military aircraft includes a light-weight radar set to detect and track high-speed targets at long distances and all altitudes down to ground level. The aircraft's computer chooses and fires appropriate missiles at the targets or fires the aircraft's internal guns. For close-range combat, the radar typically projects the target automatically onto a 'head-up display' that details all the necessary information as symbols on a TV screen at the pilot's eye level. The pilot is thus given all the information he needs to intercept and attack a hostile aircraft without taking his eyes off the target.

The head-up display unit also displays navigation information and data about the aircraft's performance so that the pilot is aware of any faults in the aircraft's system as soon as they occur. The problem with the head-up display – and the other equipment, like the attitude and heading reference set that displays information on the pitch, roll, and magnetic heading of the aircraft, and the inertial navigation set – is that to interpret the huge amount of information that the pilot is bombarded with, and to act on it swiftly enough, puts an unacceptable strain on the pilot, particularly during combat.

Military combat aircraft are rapidly becoming so difficult to fly, without having to fire and control the weapons at the same time, that the only solution will be to eliminate the pilot altogether and rely on un-manned aircraft, called remotely piloted vehicles (RPVs).

RPVs are already being used for a number of military purposes. The Israelis, for example, used RPVs extensively in the recent war in the Lebanon. Equipped with remotely controlled television cameras with zoom lenses, they were used to control artillery fire, and for real-time reconnaissance, surveillance, intelligence, damage assessment, etc. The RPVs proved to be particularly effective for real-time target acquisition along borders and seashores and over all kinds of terrain. The effectiveness of Israeli RPVs was brought home by BBC TV news pictures of the Israeli defence minister watching the battle in Beirut, while it was actually happening, on his television set in his office in Jerusalem.

One attraction of RPVs is their low operating costs. Manned aircraft are very expensive, partly, at least, because of the need to protect the pilot. Piloting a modern military combat aircraft has also become such a complicated business that it costs a great deal of money to train pilots. RPVs are, therefore, much more cost-effective than manned aircraft.

Electronic warfare

Clearly, electronics are playing a rapidly increasing role in all military activities. 'The electronic order of battle' will be the decisive order of battle in future wars.

The importance of electronic warfare puts a new premium on finding out as much as possible about the enemy's electronic systems – his radar, radio, command, control, and communications, and so on. This knowledge is used to design sensors for missiles to attack his weapon systems. In turn, the other side tries to deploy electronic countermeasures to frustrate the enemy attack. The first side then deploys counter-countermeasures, and so on in a never-ending and incredibly sophisticated electronic arms race.

Electronic warfare has spawned a whole new military activity known as electronic intelligence, or ELINT. ELINT uses world-wide land, sea, and air operations, as well as satellites, to spy on the enemy's communications, radar, and weapon systems. The main purpose of this espionage is to discover which frequencies in the electromagnetic spectrum the enemy (and other countries) uses for military purposes, and to listen in to his military, diplomatic, and domestic communications, including radio, telephone, cable traffic, and so on. ELINT is an extremely expensive, and the most secret, military activity, using the world's most sophisticated electronic apparatus. Large government establishments are involved, such as the British Communications Centre at Cheltenham. Crucial to ELINT is the use of spy satellites, like the US Ferret satellites and their Soviet counterparts. The knowledge gathered from ELINT is used to develop more effective electronic warfare devices, electronic countermeasures against enemy missiles, electronic counter-countermeasures and so on.

Conclusions

Military technology is automating warfare. Current trends will undoubtedly continue for the foreseeable future and there are no technological reasons why the conventional battlefield should not become fully automated.

The extraordinary increase in the destructive power of conventional weapons is making the battlefield so lethal that soldiers are becoming less willing to fight on it. Also, armies can no longer afford to lose large numbers of troops. As weapons become more complicated, they require operators with considerable technical skills. These people cost a lot to train and are difficult to replace. Modern battles are highly mobile, fast affairs. So much so that people on the battlefield are increasingly subject to great psychological stress.

For these physical and psychological reasons, the military establishments in the advanced countries are interested in the battlefield use of robots and unmanned vehicles. The combination of technological availability and military interest is enough to bring about automated warfare. Already, many reconnaissance, surveillance, and target-acquisition operations and many military command, control, and communications operations have been automated. Together with intelligent missiles, these are the main elements of the automated battlefield.

Chapter 2
Finding the Enemy

On 6 June, 1982, the Israeli Defence Force invaded Lebanon in a war that Israel called 'Operation Peace for Galilee'. It was a massive invasion, involving all Israel's armed services. The Israeli Air Force's campaign essentially began on 9 June with a surprise attack on the Syrian surface-to-air missile complex in the Bekaa Valley. Soon after this action there was a massive air battle between Israeli and Syrian fighters, the largest single air battle since the Second World War. At its peak, about 90 Israeli and 60 Syrian jets were dog-fighting in the combat area.

The Bekaa Valley, in central Lebanon, is a fertile plain about 40 kilometres long and 20 kilometres wide. The valley is flanked on both sides by hills, 2,000 metres high. The Israeli aircraft used to attack the Syrian positions included F–4 Phantom and F–16 strike aircraft. The Phantoms carried Shrike and Standard missiles that could home in on the radar waves emitted by the radars working with the Syrian surface-to-air missile batteries. The F–16s were armed with stand-off missiles, capable of being launched from a distance and finding their targets without further instructions from the launching aircraft, and conventional bombs. Artillery and ground-launched missiles were used to support the attacking aircraft.

The unusual feature of the Israeli raid was that it began with the use of a wave of remotely piloted vehicles (pilotless aircraft) as decoys. The Syrians fell for this ruse and attacked the drones with surface-to-air missiles. When the Syrians switched on their radars to attack the Israeli drones, the missiles carried by the Israeli aircraft were able to detect the rays emitted by the radars and home in on them.

According to an account of the operation given by American Middle East expert Benjamin S. Lambeth, the Israeli air raid was over in a mere ten minutes. The Israeli forces destroyed most of the surface-to-air missile sites. Throughout the entire attack, Israeli remotely piloted vehicles, carrying TV cameras, circled the area and provided the Israeli commander, sitting in his ground-based command post, with continuous television cover of the events. Remotely piloted vehicles had come of age.

Lambeth describes how the Israeli aircraft attacking the Syrian positions were provided with information by several E–2C spy aircraft flying up and down off the coast of Lebanon. These aeroplanes monitored the airspace to warn the Israeli aircraft of the approach of any Syrian fighters. They could pick up enemy aircraft 350 kilometres away. A Boeing 707 aircraft, packed with electronic equipment to collect electronic intelligence, monitored Syrian radars. And CH–23 helicopters carried jamming equipment to make ineffective voice and other communications between Syrian fighters and their ground controllers. So good was Israeli reconnaissance that the Israeli Air Force was able to destroy some thirty Syrian surface-to-air missile sites in seven raids and shoot down, according to Lambeth's estimates, some eighty-five Syrian MiG fighter aircraft with only two Israeli aircraft lost. The Lebanon was yet another example of the importance in modern war of good reconnaissance and intelligence.

Gathering information about the enemy forces in a timely way has always been an essential part of military operations. The success of any military strategy depends largely on good intelligence. Military intelligence can monitor an adversary's intentions before the shooting begins. Diplomacy can then be given a chance to solve any conflict by negotiation. If a war starts, reconnaissance determines the capabilities of enemy forces, their size and location, and tracks them to determine the place where they may attack.

Traditionally, intelligence has been collected by humans. Spies operated on the ground and photographs were taken from relatively low-flying aircraft. But the methods and procedures used for reconnaissance are changing dramatically. Electronic intelligence systems and computerized analysis are fast taking over from more traditional methods.

Most people are familiar with the traditional methods of reconnaissance. In the Second World War, for example, photographs of enemy forces were usually taken from an aircraft, typically a Spitfire or Mosquito. The aircraft then flew back to its base, the film was developed, and an intelligence officer studied it through a magnifying glass to assess the size and location of any enemy forces. Relevant information was then transmitted to the appropriate military commanders in the field. By the time the commanders received the information, of course, the enemy forces would, more often than not, have moved to a new location. Nowadays military intelligence is increasingly acquired in real time. The military man is realizing his dream. Events are being observed while they are actually happening. Real-time intelligence and target-acquisition capabilities are transforming warfare.

The spies are no longer mainly on the ground or in low-flying aeroplanes. They are moving further upwards into deep space and further downwards into the deep ocean. And the main source of information has become virtually the entire spectrum of electromagnetic radiation. Optical, electro-optical (TV), infra-red, radar, and acoustic Sensors are all used to collect intelligence information.

As military intelligence relies on more and more sensors, it becomes more complex. In fact, military intelligence has become so complicated that it is usually subdivided into a number of categories: communications intelligence (COMINT), signals intelligence (SIGINT), imagery (i.e. photographic or electro-optical intelligence (IMINT), and electronic intelligence (ELINT).

Reconnaissance aircraft

Currently, we are in a period of transition. Manned aircraft are still used for reconnaissance. NATO, for example, uses the SR–71 for strategic reconnaissance and the TR–71 for tactical reconnaissance. The TR–71, a long-endurance aircraft capable of flying long distances, is based at the RAF station at Alconbury in England. It is used for high-altitude surveillance and reconnaissance. The TR–71 carries the most modern high-resolution cameras, infra-red and radar systems, and sensors to detect enemy radar and radio emissions. But the intelligence is not gathered in real time. Partially to overcome this disadvantage and give the aircraft some real-time capability, the aircraft are being fitted with the Precision Location Strike System (PLSS). This is an all-weather system that will detect enemy emitters at a distance and transmit information about the location of the targets directly to various airborne and ground weapon systems assigned to attack them.

Complementing the TR–71 is the reconnaissance version of the F–4 Phantom, the RF–4C. The RF–4C is also stationed at RAF Alconbury in Cambridgeshire, and at the West German Air Base at Zweibrucken. These aircraft are designed to fly, at low altitude and high speed, over target areas and photograph them. The aircraft also carry a variety of other sensors, including radar, electronic and infra-red. Among the radar sensors is side-looking radar to detect and track enemy forces at a distance and in real time.

Because manned aircraft are becoming increasingly vulnerable to air defence systems (see Chapter 3), increasing use is being made of remotely piloted vehicles (see Chapter 5) and satellites for military reconnaissance. In fact, the most obvious military use of satellites is for reconnaissance and espionage.

Reconnaissance satellites

High-altitude methods of reconnaissance came of age in the mid-1950s with the use of the American high-flying spy-plane, the U–2. For the first time in military history, cameras on the U–2 were able to photograph such huge areas of the Earth's surface that it became feasible to photograph entire continents. And, as the quality of cameras improved, it became possible to photograph them with increasingly high resolution.

The superpowers have since then found it very difficult to hide military activities of any scale from each other. And the superpowers have been able to observe any significant military activities in all other countries, an enormous advantage in international and military affairs.

The U–2 reconnaissance plane is, like the SR–71 and TR–71, able to fly at altitudes of about twenty-five kilometres (about three times the cruising altitude of modern commercial aircraft like the Boeing 707) to take photographs. But the need for secure and real-time reconnaissance is shifting interest away from high-altitude manned aircraft to satellites. Satellites in orbit at much higher altitudes are, in most cases, able to spy more efficiently than manned aircraft.

Until recently, all reconnaissance satellites had a similar disadvantage to aircraft. They had to eject their films after photographing their targets. The films had then to be developed and analysed, operations that took significant time, though less time than that taken by aircraft. Packages of films from satellites are normally ejected over the ocean in a capsule. When the capsule gets close to the surface of the sea a parachute opens to slow the flight. An aircraft is sent to pluck the package out of the air for delivery to the intelligence officers. However, it has recently become possible to develop photographic film automatically on a satellite and then scan the film electronically, with a television camera, for example. The information on the film is sent by radio signals to ground stations that translate the electronic data back into photographs. But currently it still takes time to process satellite images – too much time to satisfy military demands for real, or near-real, time information.

With new technology, however, it will be possible to complete the whole process so rapidly that the intelligence officers can scan the photographs almost immediately after they are taken. Spy satellites will then be able routinely to inform their owners about events virtually while they are happening, and the military demand for real-time intelligence will be met.

The ability to perform real-time intelligence with satellites, equipped with optical, infra-red and radar sensors, will be much improved by the use of very high-speed integrated circuits that will considerably increase the power of computers. Satellites will be able to send data to Earth in a highly compressed form. Currently, compression techniques are used to transmit live television pictures. As Fred Guteri, a writer on military intelligence, explains:

> One such technique is to section off blocks of picture elements, or pixels, that have the same level of gray and then represent the entire block by number. Another is to hold the previous image in semiconductor memory, compare that with the current image, and then specify only those pixels that have changed. Similarly, radar satellites can transmit only moving-target images . . . Both of these techniques reduce the amount of information that must be transmitted for each image.

Real-time reconnaissance is also made easier by the use of a new technological development called charge-coupled devices (CCDs). A CCD is a semi-conductor that is sensitive to light. An array of CCDs is used in a reconnaissance satellite in place of a film. The CCDs store an electric charge in the pixels that is proportional to the intensity of light. After exposure, the charge at each pixel is read directly by computers and transmitted directly to a ground station. The use of films, and their time-consuming development, is eliminated.

Key intelligence satellites rely on infra-red and optical sensors. Of these, photographic satellites are the most common. The military use of photographic satellites dates back to the early 1960s. Since then, about 40 per cent of the military satellites launched by the superpowers have been for photographic reconnaissance, indicating the importance attached to this activity by the military. Most photographic satellites are in relatively low orbits, some 200 kilometres above the Earth.

Photographic satellites normally come in two varieties. One scans a large area of territory using a wide-angle, low-resolution camera. If the military intelligence analysts spot something that interests them on these photographs, a satellite of the second type is sent to investigate in more detail with a high-resolution camera. The photographs taken by these close-look satellites show incredible detail. Not only can individual soldiers be picked out but the photographs also show what the soldiers are doing.

Increasingly, the two types of mission – large-area surveillance and close-look operations – are being done by one satellite. The American Big Bird satellite, for example, orbitting at altitudes of between 160 and 300 kilometres, does both. Generally speaking, though, the Soviets still use different satellites for each of the two missions.

What satellite photographs can show

Not very surprisingly, information about the quality of military satellite photographs is top secret. Some information has, however, become available.

When the Space Shuttle *Columbia* was launched in April 1981, the crew noticed through their television monitor that some thermal tiles had been torn off during the launch. If too many tiles had been lost the crew would have been in danger when the spacecraft re-entered the Earth's atmosphere, because it would have become over-heated without an adequate heat shield. Although the crew could monitor the top of the shuttle, the television monitor could not scan the bottom side. To check this part of the spacecraft, the Pentagon took some photographs from Earth of the spacecraft when it was in orbit at an altitude of 100 kilometres. The photographs showed the tiles on the underside of the shuttle clearly enough to tell that none were missing. The tiles were 7.6 centimetres square. The fact that they can be seen on a photograph at a distance of 100 kilometres implies a resolution of 10 centimetres or less, if the atmosphere is reasonably clear and the weather is good. Some idea of what this resolution means in practice can be gained from an account written by American military affairs correspondent Drew Middleton in the *New York Times*, on 11 September 1983:

> The United States and NATO specialists said satellite pictures were the mainstay of visual intelligence. Pictures taken from

> satellites flying at 100 or more miles altitude after magnification have in one instance shown the bolts on the deck of a Soviet cruiser. In another picture, a man reading a newspaper on the street of a north Russian town was seen in the picture to be perusing *Pravda*. It is known to be *Pravda*, they note, because the nameplate is clearly visible in the satellite photograph.

Although military reconnaissance satellite pictures are kept strictly secret, one was unofficially obtained and published recently in a reputable defence magazine, *Jane's Defence Weekly*. It shows a Soviet nuclear-powered aircraft under construction. Even the raw photograph shows much detail. Modern analytical methods can get a great deal more information from the photograph. One such method is digital image processing (DIP). The technique is well described by David Hafmeister, an American expert on satellite technology, writing in the *Bulletin of the Atomic Scientists*:

> Photographic systems tend to blur straight lines and edges into a gray region of transition from light to dark. This blurring can be caused by the optical system, fluctuations in the air or motion. DIP methods can restore lines and edges by searching for the shortest distance through the gray transition region in the same way that we could search out the steepest and shortest path down a mountain. After a number of these steepest paths have been discovered, the computer can then determine the location of the edge or line. Once the computer has found the line, it restores the image by removing some of the data from the foothills and returning it to the mountain top.

As computer power increases, DIP techniques will improve and it will soon be possible to equip reconnaissance satellites with DIP technology. The combination of DIP and CCD technologies will greatly enhance the performance of reconnaissance satellites. The resolution of the images obtained with arrays of CCDs is reportedly as good as silver-halide based photographic films.

The new technologies will be used in sophisticated reconnaissance satellites like the American KH–11. They orbit at altitudes of between 250 and 500 kilometres and provide real-time intelligence. Currently, their pictures have resolutions of about two metres but this will improve as the new technologies are incorporated. Eventually, they will compete with today's close-look satellite pictures that have resolutions of about 10 centimetres or less but that are on films that are sent back to Earth.

Ocean surveillance and ELINT

One type of reconnaissance satellite surveys the oceans. Some ocean-surveillance satellites identify and track warships. Their mission is to keep track of all surface warships at sea at any one time. Typically, the satellites orbit at altitudes of about 800 kilometres and normally use radar to do their job. Ocean-surveillance satellites may also carry infra-red sensors. If so, they can be used to complement anti-submarine warfare activities (see Chapter 4).

Radar and infra-red data from the satellites are fed into computers with data from the various anti-submarine warfare sensors. These sensors pick up signals from enemy submarines but they also pick up many other unwanted signals that form a 'background noise'. The satellite data are used to compute partially the ocean background 'noise' signals which are then subtracted from the data from the anti-submarine warfare sensors. This improves the efficiency of anti-submarine warfare detectors.

The radars on Soviet ocean-surveillance satellites are usually powered by a small nuclear reactor carried on board the spacecraft. In 1978, one of these satellites, Cosmos 954, became famous. It fell out of orbit and scattered radioactive material when the nuclear reactor burnt up as it re-entered the Earth's atmosphere. Some of the radioactivity contaminated part of Canada, much to the annoyance of the Canadian authorities.

The most sophisticated spies in space are the electronic

surveillance satellites operated by electronic intelligence, or ELINT, agencies. These satellites carry extremely complicated electronic equipment to collect and analyse radio signals generated by the enemy's military forces as they go about their business. The signals of particular interest to ELINT officers are those arising from radio and telephone communications between military units, communications between military units and their command centres. Also of great interest are all types of radar signals – from air-defence systems, from warships, from missile-control units, from early-warning-of-attack systems, and so on.

The main mission of ELINT satellites is to locate the sources of the electronic signals generated by the enemy's military forces and determine the characteristics of the signals. This information is used to design weapons to penetrate the enemy's defences. It is also used to design electronic countermeasures to frustrate the enemy's weapons. The enemy will, of course, respond by developing electronic counter-countermeasures, and so on, in a never-ending electronic arms race. Given the importance of electronic warfare in today's battlefield, ELINT satellites are regarded by the superpowers as the most crucial of their spies in space.

American ELINT satellites include Rhyolite and Ferret satellites. The Rhyolite satellites operate in geostationary orbits, at altitudes of 36,000 kilometres above the Earth's surface. In these orbits they remain stationary over the same area of the Earth's surface and can, therefore, keep a continual watch on that area. Ferret satellites occupy orbits at altitudes of about 600 kilometres. One of their main missions is to monitor Soviet radar. Both types of satellite are used to keep track of Soviet ballistic missile tests.

ELINT satellites may have pride of place among satellites, but American Vela satellites easily hold the record in size of orbit. The purpose of Vela satellites, orbiting at an altitude of about 110,000 kilometres in really deep space, is to detect and locate nuclear explosions in the Earth's atmosphere and in outer space. Twelve Vela satellites have been launched and three are still operational.

The Americans will in future use the Integrated Operational

Nuclear Detection System (IONDS) to detect nuclear explosions in the atmosphere and in space. Optical and X-ray sensors for IONDS are being installed on NAVSTAR satellites, the new American IS-satellite navigational system.

Radar

Many reconnaissance and target-acquisition systems rely on radar, shorthand for 'radio detection and ranging'. Radar uses directed beams of radio waves to detect, locate and track moving objects and to accurately locate fixed targets. A target is detected by the reflection of the radio beam off the target back to a receiver. (The reflected beam is called the 'echo'.) The radar's transmitter and receiver are usually located together and use the same aerial.

Radar works in all weathers, day and night. Usually, the radio waves are in the microwave band although radars used for very long range operations use high-frequency (HF) and very-high-frequency (VHF) band. Radar frequencies, therefore, range from 3 million to 10,000 million hertz.

Normally, radar waves are sent out in pulses, at a pulse rate of up to many thousands of pulses a second. Most often, a pulse is not sent out until the previous pulse has travelled all the way to the target, been reflected by the target, and arrived back. The time-lag between the time at which the pulse is sent out and the time when the echo returns gives the distance of the target. The direction from which the echo returns gives the bearing of the target. The strength of the echo can indicate the size of the target.

Because of the military demand for rapid information about targets, and the huge amount of data that can be collected by radar, it is normal to process and display the information by computer. The computers may use the information about enemy forces collected by the radar to assess the threat from these forces and choose appropriate weapons to attack them. The computer can also be instructed to fire the weapons automatically and, with continuous information from the radar, to guide them to the target. The entire process from target acquisition to firing the weapons can be done in a few thousandths of a second. No human can compete with such rapid action.

The capabilities of radar are being improved by the use of over-the-horizon-backscatter and phased-array radars.

Over-the-horizon-backscatter (OTH–B) radar is not limited in range by the curvature of the Earth. It can 'see' objects several hundred kilometres away, far over the horizon. OTH–B radars considerably increase the amount of warning time of an attack and enhance the effectiveness of defence systems. The OTH–B technique is to reflect radar signals, and the returning echoes, off the ionosphere (an electrically-charged layer in the atmosphers), in a way similar to the reflection of light in a mirror. Ordinary radio signals are transmitted over long distances in the same way.

Important though OTH–B radar is, phased-array radar is even more so. The traditional concave radar dish, endlessly scanning the horizon, is now a familiar object seen, for example, at all airports. But, with the advent of phased-array radar, the rotating dish will become a thing of the past.

Phased-array radar uses a fixed flat area consisting of many small aerials instead of a single large rotating aerial. Each of the identical small aerials transmits and receives signals. Each signal is deflected from target to target electronically. The advantages of phased-array radar have been described by Eli Brookner in an article in the journal *Scientific American*. Brookner explains that a conventional radar can gather information about a large number of targets:

> Inevitably, however, there is a substantial lag between successive observations of individual targets. The update rate, at which a radar takes new readings of a target, is for most mechanically steered radars no faster than the rate at which the radar dish turns on its shaft.

Even the best military radars cannot achieve update rates better than twice a second. More often than not, military operations require continuous observation of each of a large number of targets. If rotating-dish radars are used, this would require a number of radars, each assigned to one or a small number of the targets. As Brookner explains:

> Today a single phased-array radar can do what previously might have required a battery of mechanically steered dishes. To give a concrete example, a phased array code-named the COBRA DANE, which observes Soviet ballistic-missile tests from its site on the edge of the Bering Sea, can follow hundreds of targets scattered through a volume of space spanning 120 degrees in azimuth and about 80 degrees in elevation. In effect it watches them simultaneously, electronically redirecting its beam from target to target in a matter of microseconds (millionths of a second).

Incidentally, the COBRA DANE phased-array radar contains 15,360 aerial elements in an array about 30 metres in diameter.

Phased-array radars in Florida and Massachusetts can detect objects in the atmosphere and space at distances of up to 5,000 kilometres, and determine their size and shape. And two phased-array radars, called PAVE PAWS, are operating on the east and west coasts of the USA where they continuously monitor the areas where Soviet strategic nuclear submarines operate to give early warning of attack by submarine-launched ballistic missiles. Each 5,500-kilometre range PAVE PAWS radar, that is also used to track satellites, has 1,792 aerials on each of two faces. Each face is 31 metres wide and can scan 120 degrees of azimuth. The faces are at an angle of 120 degrees to each other so that the radar has a 240-degree field of view. Two more PAVE PAWS radars are being built in Georgia and Texas. In addition, the Americans operate the Perimeter Acquisition Radar Characterization System (PARCS), a phased-array radar in North Dakota, and are building phased-array radars at Thule, Greenland, and at Fylingdales, England.

The Americans are also deploying a network of over-the-horizon radars on the east and west coasts of the USA. These radars will be able to identify and track enemy aircraft hundreds of kilometres away at any altitude. New radars to track bombers are also being deployed in Alaska and northern Canada.

All these phased-array and over-the-horizon radars will form a network of reconnaissance radars to monitor objects in the atmosphere and space – including missiles, aircraft and satellites – for the USA and its NATO allies.

Airborne warning and control system aircraft

Tactical reconnaissance of the airspace above the battlefield is also provided by Airborne Warning and Control System (AWACS) aircraft. These exceedingly expensive aircraft carry the most up-to-date radars and computers. The most recent American AWACS aircraft, the E–3A, cost $1,500 million in research and development and each aircraft costs more than $100 million to produce, about six times the cost of the most sophisticated US fighter aircraft.

The US operates a fleet of thirty-four E–3As to monitor the airspace of North America, Central Europe and the Greenland–UK gap. The USSR also operates a fleet of AWACS aircraft, based on the Tu–126 Moss, but the Soviet AWACS aircraft are much less sophisticated than their American counterparts. In particular, the American AWACS are equipped with advanced electronic countermeasures giving substantial resistance to enemy jamming.

The US Navy also operates AWACS aircraft – E–2Cs – from its carriers to provide airborne early warning and command and control support for air defence and sea control missions for carrier battle groups. A notable feature of the E–2C aircraft is the 7.3 metre diameter saucer-shaped rotating dome fixed above the fuselage which carries the radar aerial. Each E–2C costs about $60 million.

The purpose of an AWACS aircraft is to detect and identify all suspicious flying objects in the airspace it monitors soon enough to ensure interception and, in addition, to provide extensive command and control facilities for all friendly aircraft within its range – including interceptors, transport aircraft, reconnaissance aircraft and so on. The identification and tracking of hostile aircraft and the control of friendly ones – at long range, at high and low altitudes, in all weathers and over land and water – are achieved by the use of over-the-horizon, look-down surveillance radars, high-speed computers and multi-purpose display units.

The E–3A is a modified Boeing 707. Its liquid-cooled radar aerials, housed in a rotating dome fixed over the fuselage, can scan the airspace from ground level to stratospheric altitudes. The E–3A patrols for up to eleven hours, or for more than twenty hours with air refuelling. It is, therefore, provided with exceptionally accurate navigation systems. Typically, about twelve AWACS specialists would be carried in each aircraft, in addition to the crew of four.

Many experts regard AWACS as an enormously expensive white elephant, arguing that the aircraft would be a prime target in any war and would soon be shot down, probably within minutes. Critics also say that the aircraft is inadequately protected against electronic countermeasures, given that it will operate in an environment saturated with electronic countermeasures.

AWACS advocates argue that AWACS radars can, on the contrary, resist countermeasures better than ground-based surveillance radar can, and that the aircraft would survive long enough under attack to fulfil a worthwhile role. They emphasize the value of an early-warning period against surprise attack, even if this period is a short one. Any warning, in other words, is better than no warning.

The Pentagon seems to agree more with the critics. It says that new wide-area ground-based sensors, such as over-the-horizon radars, 'have a higher priority and greater long-term potential'.

Joint Stars

The Joint Surveillance and Target Attack Radar System (JSTARS) is, like AWACS, an airborne radar system used for target acquisition. But, whereas AWACS is used to detect aircraft, JSTARS is designed to locate, identify and track moving armoured vehicles, particularly tanks.

JSTARS is designed to operate effectively in all weathers, at long range and in real time. The 'side-looking' radar equipment is able to locate tanks at ranges of 150 kilometres and more. The aircraft carrying JSTARS would normally circule some distance from the battlefield. The aircraft would, therefore, be less vulnerable to enemy air defences than it would be if it had to cross the lines.

The real-time information about targets collected by JSTARS is passed to a control station. Computers in the control station analyse the information and may choose and fire appropriate weapons, normally missiles, to attack the targets. The JSTARS radar guides the weapons during flight to the target. Currently, human operators in the control station are involved in the control of weapons called for by JSTARS, but there is no reason why the entire control process cannot be computerized. JSTARS could also be carried by remotely piloted vehicles so that no humans need be involved at all in the operations leading to the acquisition and destruction of targets.

The JSTARS system can be complemented by the Precision Location Strike System (PLSS). Both systems can be carried in the same aircraft. PLSS will detect and locate enemy systems that are emitting electromagnetic radiation (radio waves, radar, and so on). Like JSTARS, PLSS transmits the locations, in real time, to weapon systems so that the targets can be quickly attacked.

JSTARS and PLSS are specially equipped with electronic counter-countermeasures so that they can operate effectively even when the enemy is doing his best to jam the equipment. Details of anti-jamming measures are, for obvious reasons, very tightly held secrets.

Stealth

As we have seen, radar is used by many reconnaissance and target-acquisition systems for a variety of functions including early warning of attack, locating, identifying, and tracking targets, and guiding missiles to their targets. Radar is the part of the electromagnetic spectrum most used by the military.

Developments in radar and other electronic warfare technologies are part of what Daniel Deudney, an American expert in information technology, has called the 'transparency revolution'. Advances in these military technologies are, he says, often 'rooted in the application of information technologies to warfighting. Advances in information technology – sensors, communications and data processing – have created a rudimentary planetary nervous system, fragments of a planetary cybernetic.' Deudney argues that central to the transparency revolution is the militarization of the electromagnetic spectrum. This is, as he puts it, the militarization of yet another 'natural feature of the planet lying beyond the effective sovereignty of the nation state'. Deudney points out that the worldwide military reconnaissance, target acquisition, and command, control, and communications systems set up by the superpowers have formed a 'planetary-scale web of electronic intelligence that alters the potency of weapons as well as the incentives for using them'. The emphasis in warfare is shifting away from the destructive power of weapons to the ability to detect and target the enemy's forces and 'hide and communicate with one's own'.

Deudney believes that future warfare will no longer be the 'traditional struggle between offensive and defensive military forces' but rather 'a competition between the visible and the hidden – between transparency and stealth' with the most important element in transparency technologies being the co-ordinated and accurate targeting of weapons.

Given the importance of radar in this process, it is hardly surprising that large resources are being spent on means of frustrating radar. James Pellien and Peter Williams, two electronic warfare experts, have listed five threats to radar: the reduction of the radar cross-section of the target; high-powered jammers; anti-radiation missiles; terrain-following aircraft; and remotely piloted vehicles.

Anti-radiation missiles home on the radiation transmitted by an enemy radar; the radar itself provides the beam down which the missile travels to destroy the radar. Radar contains the seed of its own destruction. This is its major weakness. The potential of anti-radiation missiles was dramatically demonstrated by the famous Israeli attack on Syrian air defences in the Bekaa Valley on 10 June 1982. Pellien and Williams describe the operation as follows:

> The Bekaa Valley was heavily defended by some 40 to 50 SA–6 and SA–8 Soviet built, Syrian manned air defence systems. The Israelis first of all destroyed the radars associated with the missile systems and then, having rendered them unable to fire their missiles, destroyed them all by ground and air attack without loss. The key to this classic operation was a series of deception attacks that made the Syrians switch on their radars, the acquisition of the radar characteristics by RPVs and then the destruction of the radars by anti-radiation missiles, using the acquired radar characteristics for homing.

The novel use of RPVs in the operation is further discussed in Chapter 5.

The Israeli attack in the Bekaa Valley was in a sense a tit-for-tat. In the Yom Kippur war, in a less well known operation, two Israeli radar stations were destroyed by Soviet-supplied cruise missiles that homed in on the emissions from the stations.

Aircraft use a number of strategies to avoid detection by enemy radar. One is to use radar, called terrain-following radar, carried by the aircraft that enables it to 'hug' the ground. The aircraft flies so low and fast that radar-controlled air defence systems have very little time to engage it. By the time the radars have 'seen' the aircraft and have selected a surface-to-air missile to fire at it, the aircraft has gone.

The American B–1B strategic bomber, for example, is a terrain-following aircraft. The aircraft carries a phased-array radar, the Offensive Radar System, and advanced computers that will allow it to fly automatically about 70 metres above the ground at nearly sonic speeds. Air defence systems would then have just a few seconds to detect and fire at the aircraft. The 186-ton B–1B carries the world's most advanced defensive electronics (avionics), including electronic countermeasure equipment. It is a very expensive aircraft – the US Air Force is buying 100 B–1Bs at a cost of $210 million each.

The radar cross-section of the B–1B is ten times less than that of its predecessor, the B–52. The reduction is achieved by the shape of the aircraft and the nature of its surfaces. These reduce the amount of energy in a radar beam reflected from the aircraft. Rounded surfaces, for example, reflect less energy than angular shapes. The radar echo is, therefore, decreased and harder for the radar's receiver to separate from the 'noise' level. ('Noise' is produced by false echoes from the ground and spurious signals produced in the radar equipment.)

Much research is going on to reduce the radar cross-section of aircraft even more. The aim is to produce a bomber with a radar cross-section ten times less than that of the B–1B. Aircraft with very low radar cross-sections are called 'stealth' aircraft. Stealth technologies are secret and very little has been written in the open literature about them.

It seems that stealth is achieved by a combination of shaping to reduce angular shapes, new radar-absorbing materials, and advanced electronic countermeasures. Bill Sweetman, an aerospace expert, says that stealth aircraft designers aim at buried engines, extensive wing-to-body blending to eliminate vertical planes, and generally minimized side and front profiles. According to Sweetman, radar-absorbing materials include carbon compoundss and thermoplastic compounds. He also believes that advances in composite and plastics technologies will provide a new and extensive range of radar-absorbing materials.

Stealth aircraft will, of course, never become invisible. They are also likely to be extremely expensive. All in all, for many, if not all, important military purposes, remotely piloted vehicles are far more cost-effective than manned aircraft. Stealth technologies are a last-ditch effort to keep pilots in business.

Chapter 3
Weapons for the Automated Battlefield

The year is 1995. Two industrialized countries are at war. One side decides to invade and occupy the other. Its tanks, in regular columns, approach the no man's land at the border between the two combatants.

Suddenly, a swarm of small pilotless aircraft appears above them. The aircraft are not much bigger than the larger of the radio-controlled aircraft you see in toyshops. The tank gunners try to aim their machine guns at the enemy aircraft. But the drones fly fast, in zig-zag patterns and at low altitudes. Hardly any are shot down before they fire their anti-tank warheads. The warheads look like molten chunks of metal, about twelve centimetres across.

Soon, about half the tanks are knocked out. The warheads fired by the drones easily pierce the armour. They seem to know exactly which parts of the tanks are the weakest. They attack the turrets and engine covers from above with deadly accuracy. The warheads can think for themselves.

Small computers carried by the warheads do the thinking. Sensors in the warhead provide information for the computer, just as our senses collect information for our brains. Actually, the warhead's sensors are sensitive to more types of radiation than human senses. We can hear sound waves, see visible light waves and smell and taste chemicals. The warheads are also sensitive to sound and light waves, and they can also detect the infra-red rays given off by the hot engines and exhausts of the tanks and the millimetre-wave radiation (radiation having very short wavelengths, of a few millimetres) given off by the tanks. Because millimetre-waves are unique to tanks they are an excellent way of detecting tanks and discriminating between them and other vehicles.

A hit on a tank by one of the warheads fired by the drones is lethal to the crew inside – red-hot pieces of metal ricochet around the inside of the tank. Hot, suffocating gases spread rapidly through the tank. Very few survivors crawl out of the burning wrecks.

The remaining tanks regroup and advance cautiously. This time, they use what cover they can. But, without warning, the three tanks at the front of the formation burst into flames. The way is blocked. All the tanks must stop.

Almost immediately three more lead tanks burst into flames. What is killing the tanks so effectively? The crews in the tanks close to those destroyed see several small enemy vehicles coming towards them. The vehicles are unfamiliar. One of them fires two missiles. Another tank bursts into flames. And then another.

The gunners that can see the enemy vehicles fire at them. The tanks' computer-controlled guns are very accurate. The enemy vehicles quickly go up in flames. A tank commander sends out a man to look at the enemy vehicles. They are robot-driven.

With no humans to protect, the enemy anti-tank vehicles need no armour. They are relatively simple vehicles, except for their computers, which contain the very latest circuitry. They each have six large wheels, with tyres rather than tracks, almost as big as farm-tractor wheels.

The vehicles look ungainly but they can travel fast over rough ground, much faster than the sixty-ton monsters they are attacking. Against such a mobile enemy the tanks are sitting ducks. Wisely, they decide to head for the cover of bushes or small hills and wait for darkness before moving closer to the battlefield.

When darkness falls, the surviving tanks leave their cover and head once more for the battlefield. This time they take no chances. Turrets tightly shut, they head for a nearby road. Travelling at full speed down the road the crews feel more confident. But their confidence is misplaced. Darkness is inadequate cover. Without warning, small missiles silently attack out of the night.

The new missiles are even more intelligent than the earlier ones. Each missile hovers momentarily above the tanks, selects one and attacks it by firing a high-speed projectile – like a molten metal plug – at the relatively lightly armoured turrets. The missiles are very selective. No missile attacks a tank that has been selected for attack by another missile. Each missile knows what its partners have in mind. A tank is attacked twice only if the first missile fails to stop it.

These missiles are fired from ranges of about thirty or forty kilometres, far beyond the range of the tanks' guns. A rocket launcher fires a salvo of twelve missiles, at a rate of about one every three seconds. Each missile is about a metre long and ten centimetres in diameter. The guidance system, using a number of different types of sensors, including an infra-red sensor, a TV camera and a micro-wave sensor, is carried in the nose of the missile. The sensors can seek out and identify tanks within an area of 500 square metres.

Behind the guidance system is the computer and behind the computer is the armour-piercing warhead. And at the rear of the missile are stabilizing fins which unfold after the missile leaves the launcher tube.

There are not enough missiles to destroy all the tanks. Three survive. Their morale completely shattered, the crews decide to retreat.

The men in the tanks are intelligent. Tanks have become so complex that highly skilled crews are needed to operate them. These men can calculate how small their chances of survival are. The battle is far too lethal for them. Old-style infantrymen may be prepared to face low odds, but not modern tankmen. Good education and high standards of living make people 'soft'.

None of the retreating tanks make it back to friendly territory. More robot-driven enemy anti-tank vehicles lurk in their path. The missiles fired by these vehicles can also see in the dark.

None of the tanks even reach the border. Had they done so, they would have faced a formidable minefield. The mines can detect and identify tanks, discriminating between tanks and other vehicles. They are extremely difficult to detect. Such minefields cannot be cleared. The mines explode only when a tank is above them. The destruction of the tank is virtually guaranteed.

Any tank that stops in the minefield, or is lucky enough to escape from it, would be destroyed by mortars, fired from distances of about six kilometres and hidden behind hills or woods. A microwave sensor in the nose of the mortar switches on as it begins its descent. The sensor can detect tanks within an area of 400 square metres and home in on the tank's vulnerable turret.

The engagement is expensive for the side losing the tanks. The enemy has destroyed a large number of tanks before they could reach the battlefield at a cost fifteen times cheaper than the cost of the tanks.

Why are the tankmen taken by surprise? Mainly because their army is run by generals who have had their careers in the tank corps. Men who refuse to listen to the advice of the technologists. Men who cannot face the fact that tanks may be obsolete.

As recent conflicts in, for example, the Falkland Islands and the Middle East have shown, modern warfare relies more and more on missiles. Armoured vehicles, aircraft and warships are the main platforms for firing missiles. In turn, missiles are used to destroy these weapon platforms.

A main aim of current military technology is to develop missiles that are effective under all the adverse conditions under which battles are fought: when the battlefield is covered with thick smoke and dust; in adverse weather conditions, including heavy rain and snow, thick fog and haze; and against enemy countermeasures, including chaff (metal strips) to confuse radars, electronic jamming devices, and decoys.

Another main aim of military technology is to perfect autonomous ('fire-and-forget') missiles.

Today's most widely deployed fire-and-forget weapon is the Maverick, an air-to-surface missile produced in the USA and carried on the combat aircraft of many of the world's air forces. The Maverick is about 2.5 metres long, about 30 centimetres in diameter, with a span of about 71 centimetres, and a weight of 210 kilograms. Its solid-fuel rocket motor gives the missile a range of about 22 kilometres. Production of Maverick missiles began in 1968; since then more than 25,000 have been produced. Various versions of the missile have been developed using television, infra-red, or television-plus-laser guidance systems.

In the earlier TV-guided model a small TV camera in the nose of the Maverick transmits a picture, while the missile is in flight, back to a TV screen in the cockpit of the aircraft that fired the missile. The pilot selects the target visually and centres the cross-wires on the TV display on the target. This action by the pilot locks the missile onto the target; the Maverick then automatically homes onto the target. The pilot is free to leave the area, either to attack other targets or to go home. The early TV-guided Mavericks do not work well in bad weather or at night. The infra-red version performs much better in these conditions. the missile uses sensors that detect variations in the amount of heat given off by various objects and is particularly effective against tanks.

The laser Maverick is a highly accurate version of the air-to-surface missile; its sensor guides it towards laser light that has been beamed onto the target by an operator on the ground or in another aircraft. The missile travels down the laser beam to its target. The latest infra-red and laser Mavericks cost about $140,000 each.

The performance of autonomous missiles will continue to be improved, particularly by the use of sensors operating on frequencies in the far infra-red and on radio frequencies. Specifically, there is considerable interest in the use of millimetre-waves for missile guidance; in fact, millimetre-wave technology is likely to become the dominant technology in missile-guidance systems for the automated battlefield.

Millimetre waves have frequencies of about 30,000 megacycles per second and are able to penetrate the atmosphere even when it is polluted with a great deal of dust, smoke, fog, and so on. Ultra-smart and brilliant missiles will probably rely mainly on millimetre waves. Future battles fought with these intelligent missiles will go on in virtually all weathers and under all conditions.

An example of a microelectronic development to increase the IQ of missiles is the very-high-speed integrated circuit (VHSIC).

Today's microchips, although only a few square centimetres in cross-section, have powers equivalent to those of room-sized computers of twenty years ago. In this period, the number of components per circuit has increased a thousandfold. The Pentagon's VHSIC programme aims at fabricating integrated electronic circuits 50 to 100 times more compact than existing microcircuits. These devices will operate at least twice as fast as current devices, and will require less maintenance and be more reliable. New lithographic techniques will make possible the construction of even smaller microprocessors. Eventually, computers with dimensions on the molecular scale will be made.

The military applications of these ultra-small devices will include the development of increasingly intelligent missiles able to detect and attack specific targets with very high discrimination against background noise; signal processors capable of hundreds of millions of operations a second; and systems to frustrate enemy electronic countermeasures. A specific example of a future application of very-high-speed programmable signal microprocessors is the recognition of human speech in real time so that an operator can activate systems by voice and thereby increase the number of operations he can cope with.

How missiles evolve: anti-tank missiles

The evolution of missiles is well illustrated by considering, as an example, the development of anti-tank missiles. A typical anti-tank missile now in operation is the US TOW. TOW stands for Tube-launched, Optically tracked, Wire-guided. The missile became operational in 1970 and was used extensively in the Vietnam War and in wars in the Middle East.

TOW missiles are currently to be found in the arsenals of many of the world's armies, having been exported by the USA to Canada, Denmark, West Germany, Greece, Iran, Israel, Italy, Jordan, South Korea, Kuwait, Lebanon, Luxemburg, Morocco, the Netherlands, Norway, Oman, Pakistan, Saudi Arabia, Spain, Sweden, Turkey, the UK, Yugoslavia, and a number of other countries. Altogether some 400,000 TOWs have been produced; new ones are being made at a rate of almost 1,000 a month. TOW is undoubtedly the most numerous guided missile so far produced.

TOW is so popular because it destroys tanks very efficiently and because it is relatively cheap; each missile costs about $15,000. It can destroy with a high probability a main battle tank costing $3 million or more, at a range of nearly four kilometres. Even allowing for a contribution to the cost of the vehicle carrying the anti-tank missile, the TOW is still a very cost-effective weapon.

TOW is carried on a jeep or an armoured car or in a helicopter. The launcher, which looks rather like a gun barrel, can also be mounted on a tripod on the ground so that the missile can be fired from the ground instead of a vehicle. The missile, 116 centimetres long, 15 centimetres in diameter and weighing 19 kilograms, is loaded into the launcher like an ordinary round of ammunition. The system is surprisingly light – the total weight is about 80 kilograms.

The operator of the missile looks through an optical sight. When he sights an enemy tank he keeps his viewer on the tank and presses the trigger. The launch rocket motor propels the missile from the launch tube. This motor burns out, however, just before the missile leaves the launcher. After the missile has flown a few metres, a second motor ignites to propel the missile during its flight. This rather complicated procedure is to protect the operator from exhaust gases and propellant particles.

As soon as the missile emerges from the tube small wings unfold and an infra-red flare in the missile's tail ignites. The operator's viewing system tracks the flare and a sensor in the viewer measures the position of the flare, relative to the line-of-sight between the operator and the tank. In this way, deviations from the line-of-sight are detected.

If the missile strays from the line-of-sight, a computer sends a command to the missile to correct its flight path. Signals from the sensor are sent along two fine wires attached to the missile; the wires unwind from bobbins on the missile as it flies. The signals sent along the wires operate control fins attached to the tail of the missile and guide the missile to the enemy tank. So long as the operator keeps the cross-hairs in his viewer on the target the missile will follow the line-of-sight automatically.

A modern TOW carries a 150-millimetre warhead capable of penetrating the armour of all existing tanks. The latest version of the missile (called TOW 2) has been provided with a heat source as well as the infra-red (Xenon) flare. The second beacon improves tracking at night and when the battlefield is obscured by smoke or dust or bad weather.

The TOW missile is one of an international family of similar anti-tank missiles. Others include the Soviet Sagger (used by the North Vietnamese against American tanks during the Vietnam War and by the Syrians against Israeli tanks in the 1973 Middle East war); the British Swingfire and Vigilant systems; the West German Cobra and Mamba missiles; and the European Milan and HOT (High sub-sconic, Optically tracked, Tube-launched) systems.

But anti-tank missiles like these have their limitations. One is that they must follow the line-of-sight. The projectile can, therefore, hit its target only if the tank stays in the operators's sight throughout the flight of the missile. TOW, for example, travels at a maximum speed of about 1,000 kilometres an hour but still takes fifteen seconds or so to travel three kilometres. There is a good chance that the tank will disappear behind cover during these fifteen seconds, particularly if the crew suspects that anti-tank missile launchers are in the vicinity. Another problem with TOW-type missiles is that the operator is exposed to enemy fire while the missile is in flight and, if the enemy observes the spot where the missile is fired from, the operator's position is known. If the enemy discovers the operator's position the chance of his hitting the target is, to say the least, not high. Yet another problem is that missiles guided by infra-red can be easily decoyed by, for example, an infra-red flare fired by the enemy. The missile may also be deflected by, for example, a burning vehicle in the vicinity of the target.

Many of the disadvantages of wire-guided missiles are eliminated in the new generation of anti-tank missiles, just entering the arsenals. An example of the new type of anti-tank missile is the US Hellfire (Heliborne-Launched Fire and Forget) missile. Hellfire is designed to take one of a number of different sensors, including a semi-active laser homing device. Using the laser sensor, the missile is guided to an enemy tank by a laser beam reflected off the tank. An operator 'illuminates' an enemy tank by projecting a laser beam onto it. The sensor detects the reflected laser light and the missile, as it were, rides down the beam to the target.

The person operating the laser designator does not need to be in the vehicle firing the missile. Normally, Hellfire missiles are carried on helicopters. The laser-beam operator can be on the ground or in, say, another helicopter. This gives the Hellfire a big advantage over systems like TOW; the people operating Hellfires are much less vulnerable than TOW operators because the person guiding the missile can be separated from it. The enemy cannot then pinpoint the operator by observing the spot from which the missile was fired.

Other sensors planned for Hellfire include infra-red sensors and TV seekers. With an infra-red sensor, Hellfire would be truly independent of any operator. Once the missile was fired it would seek out and attack its target by homing in on the infra-red radiation given off by the target.

But even with laser designation, currently the normal method of guiding a Hellfire, neither the helicopter firing the missile nor the laser designator are in much danger. Even though there must be a clear line-of-sight between the designator and the target during the flight time of the missile, there does not need to be a line-of-sight between the helicopter and the enemy tank. The long range, six kilometres, of the missile, its high cruise speed, and the ability of the helicopter to launch the missile from behind cover, all help to reduce the risk. The favourite method of firing Hellfire missiles is to use a laser designator on the ground, with the helicopter hidden behind cover. When the missile is fired it climbs to a pre-selected point where it can 'see' the laser light reflected from an enemy tank. Alternatively, the helicopter can fire its missiles from positions out of range of enemy defences.

Hellfires are mainly carried on the US AH–64 Advanced Attack Helicopter, called the Apache; each helicopter carries up to sixteen missiles. The missiles can be launched in salvo. Several laser designators are used simultaneously, each using a different pulse frequency. Each missile responds to one of the frequencies and is guided individually to a specific target. A Hellfire carries a 175-millimetre warhead, effective against all existing battle tanks. A small squadron of Apache helicopters armed with Hellfires is, therefore, a formidable anti-tank force, able to knock out a relatively large attacking tank force. Like many other types of modern attack helicopters, Apache helicopters are able to operate effectively during day or night and in all kinds of weather, anywhere in the world.

Although a Hellfire missile is more expensive than, say, a TOW, it is still relatively cheap, at about $40,000 a missile, when one remembers that it stands an excellent chance of destroying a $3 million tank.

The third generation of anti-tank missiles, now under development to follow the Hellfire type, will use 'sub-munitions' so that one missile warhead can carry a number of smaller missiles (the sub-munitions), each able to separately attack an enemy tank. Warheads of this type will be carried by, for example, the Standoff Tactical Missile, now being developed by American military scientists. This missile will be able to dispense sub-munitions against targets deep behind enemy lines.

The new warhead is designed to engage in a short time many enemy armoured vehicles spread over a large area. The system operates with side-looking radar carried in an aircraft or a remotely-piloted vehicle. First, the radar seeks out and tracks moving targets, like tanks, which may be far inside enemy territory. When a formation of enemy armoured vehicles is detected, a missile is launched from the ground by a signal sent from the aircraft carrying the radar. The radar also guides the missile into the air above the enemy armoured vehicles. Small sub-munitions are then released by the main missile to attack the enemy forces.

Each sub-munition is a smart missile. It can scan the target area with its own sensor and seek out, for example, a tank. When it has found one, the sensor guides the sub-munition down towards the turret of the tank. This type of guidance, in which a sensor scans the target area at the end of the warhead's flight and locks onto the target, is called terminal guidance. A warhead with terminal guidance can be guided to its target with great accuracy.

As the sub-munition approaches the tank it is attacking it fires a high-speed projectile into the turret. The turret is chosen for attack because it is the weakest part of the tank (it is the weakest part because it has to rotate rapidly carrying a heavy gun and, therefore, cannot carry such heavy armour as the rest of the tank). If each warhead carried twenty sub-munitions, a typical number, and it took, on average, two sub-munitions to destroy a main battle tank, then each main missile could destroy ten tanks – a formidable anti-tank weapon.

An example of the sort of sub-munition being developed for the new Standoff Tactical Missile is the Skeet. The Skeet is essentially a smart bomblet, weighing about 2 kilograms and carrying a 0.5 kilogram warhead. The bomblet wobbles as it falls to enable the infra-red sensor in its nose to scan an area of terrain under it. If the Skeet 'sees' an armoured vehicle it fires its warhead at it. If no armoured target is detected, Skeet is programmed to explode anyway, scattering a large number of metal fragments to destroy soft-topped vehicles and to kill people in the area.

The new American Standoff Tactical Missile will be a long-range missile, to be used for attacking enemy armoured forces, and other targets, at distances of up to 300 or 400 kilometres. A short-range anti-tank weapon under development is a new version of the Multiple Launch Rocket System (MLRS).

The version of MLRS now deployed by NATO armies fires a salvo of twelve rockets in less than a minute over a distance of up to thirty kilometres. Each rocket carries unguided grenade-like munitions which explode into many high-speed metal fragments, capable of wounding people very seriously, often fatally. As described earlier (page 3), the lethality of one MLRS salvo is comparable with that of a small nuclear weapon and shows how destructive conventional weapons have become.

Two new versions of the MLRS are being developed. One will fire rockets containing anti-tank mines. The other will fire a salvo of six smart sub-munitions. The sub-munitions will probably be gliding ones, able to glide towards the target area. Each sub-munition would use its sensor to seek out and lock onto a suitable target, such as a tank. Once it has selected its target the sub-munition would attack it in a guided dive. The plan is to provide the gliding sub-munition with a millimetre-wave sensor so that it will be effective in virtually all battlefield conditions.

A number of other smart anti-tank weapons are under development or being deployed. The Copperhead 155-millimetre artillery shell, for example, uses semi-active laser guidance. During the 1980s, the US Army plans to buy at least 30,000 rounds of Copperhead. SADARM, Sense and Destroy Armor, will be fired from 203-millimetre howitzers. After travelling thirty kilometres or so, each SADARM munition will release three sub-munitions equipped with terminal guidance systems homing in on microwaves emitted by enemy tanks. The American company Voight is developing a rocket-powered missile which relies on a very high speed, about 5,500 kilometres an hour, to penetrate the armour of tanks. At this speed the missile has a very large kinetic energy, enough, it is claimed, to punch grapefruit-sized holes in enemy tanks. The missile travels so fast as to be undetectable; it will be guided by laser beams.

In summary, an effective anti-tank defence would consist of anti-tank obstacles and smart anti-tank mines to slow down advancing enemy armour, and intelligent anti-tank artilley warheads and anti-tank missiles of various ranges to destroy the armoured vehicles. As we have seen, warheads fitted with terminal guidance have an excellent chance of hitting their targets even at long ranges. Autonomous missiles carrying warheads with terminally-guided smart sub-munitions are the weapons of the future. Against them, tanks will be extremely vulnerable.

An invading tank force would find it very difficult to penetrate an area of territory defended with modern weapons. As anti-tank weapons become smarter, the difficulty of invading with tanks will increase.

Anti-aircraft missiles

Aircraft, like tanks, are becoming more and more vulnerable to successive generations of increasingly intelligent missiles. Air defences are already difficult to penetrate. According to NATO estimates, in an air attack on a heavily defended area in Warsaw Pact territory, such as a main military air base, at least half of the attacking aircraft would be lost. Given the high cost of modern manned combat aircraft, such attrition rates are clearly totally unacceptable.

In a modern anti-aircraft system a number of attacking aircraft are engaged simultaneously. Modern radars allow air defences to track many enemy aircraft at the same time, no matter at what altitudes the aircraft are flying, and simultaneously to guide many surface-to-air missiles to their targets. The phased-array radar used by, for example, the American Patriot anti-aircraft missile system gives early warning of an air attack, tracks the hostile aircraft, fires surface-to-air missiles at the appropriate moments, and guides the missiles accurately to their targets. Most of these tasks are automated. The central computer in the Patriot system analyses the data from the radars. At the right moments the computer fires missiles and controls the interceptions. The radar continuously monitors the situation, tracking both the target aircraft and the intercepting missiles. Each missile carries a sensor that is aimed towards the enemy aircraft by the ground radar and that guides the missile to the target.

Patriot is a considerable advance on earlier anti-aircraft missiles, such as Nike/Hercules and Hawk. Whereas units of the latter missiles can identify, track and engage just one enemy aircraft at a time, a Patriot battery can keep track of a hundred aircraft and engage nine of them with nine separate missiles at the same time. The earlier missiles, which are still in service, are, however, effective.

Hawk (meaning Homing-All-the-Way-Killer) missiles were used in combat in the Vietnam War and in Middle East wars and destroyed about 96 per cent of the aircraft they were fired at. Hawk is equipped with a number of radars and sophisticated computers to locate and track enemy aircraft. It can engage these aircraft at altitudes down to about thirty metres.

A Patriot battery consists of a fire control station, eight launchers (each having four missiles), a maintenance section, and a decoy unit to protect the battery against enemy missiles homing in on the radiation emitted by the Patriot's radar equipment. Patriots are usually deployed in battalions, each battalion consisting of six fire units plus the battalion command post. A fire unit consists of a fire control station, eight missile launchers and a radar set. The battalion is a formidable air defence unit, able to simultaneously track 600 enemy aircraft and distribute them to the batteries.

The Patriot system requires considerably fewer components than other air defence systems. Hawk, for example, requires five radars, plus two main vehicles, plus supply vehicles for a battery of six launchers, each having three missiles. Patriot, however, requires just a single radar and one command vehicle for a battery of eight launchers, each having four missiles.

Patriot is a relatively large missile, over 5 metres long with a launch weight of nearly 1,000 kilograms. The warhead is a conventional explosive, probably containing about 75 kilograms of explosive. The missile reaches a speed equal to six times the speed of sound, and can engage enemy aircraft at distances of about 70 kilometres and at altitudes up to 24 kilometres.

The Patriot system is largely automated, the only manned equipment being the central control station containing the central computer. Only twelve people are required to operate a Patriot battery. The US Army believes that Patriot is a very effective anti-aircraft missile, and currently plans to procure 103 fire units and 6,200 missiles to set up 81 batteries. Of these, 54 are planned for deployment in Central Europe by the early 1990s.

Patriot missiles cost about $1 million each. This sounds a lot but it should be remembered that combat aircraft cost much more. A modern strategic bomber, for example, costs well over $200 million, and a multi-role combat aircraft like the Tornado costs about $30 million.

Patriot missiles are generally meant for use against high-flying aircraft. For defence against low-flying aircraft smaller surface-to-air missiles are normally used. This family of missiles includes the West German Roland, the Swedish Bofors RB–70, the British Blowpipe and Rapier, the French Crotale, the Soviet

SA–6 and SA–8, the American Stinger and Chaparral, and so on.

Some of these missiles, such as the RB–70, the Blowpipe and the Stinger, are portable, carried and fired by one man. The Blowpipe, for example, is less than 1.5 metres long with a diameter of about 7.5 centimetres and weighing just 11 kilograms. The missile carries a 2.2 kilogram high-explosive shaped-charge warhead and is effective against low-flying aircraft over distances of up to three kilometres. When the operator sees an aircraft approaching he finds out if it is hostile or friendly by an 'identification friend or foe' (IFF) system. If the aircraft does not respond to the challenge by giving the correct coded reply, he assumes that it is an enemy aircraft and fires the missile out of its tubular container. An infra-red sensor attached to the container detects the flare of the missile and centres it on the line-of-sight. Blowpipe is guided to the enemy aircraft by command signals from a thumb controller clipped to the container.

The RB–70 is an interesting missile because it uses a laser beam for guidance. The laser beam is very hard to jam by electronic or other countermeasures. This gives the RB–70 a distinct advantage over most other portable surface-to-air missiles which are susceptible to countermeasures.

The British Rapier anti-aircraft missile is more sophisticated than the portable hand-held types. In its original form, the Rapier system is carried in two vehicles, such as Land Rovers. The first vehicle carries four missiles loaded into a launcher, ready to fire. The second vehicle carries nine missiles for reloading. In later versions, the Rapier system is carried in an armoured vehicle.

The Rapier launcher is equipped with a surveillance radar to detect aircraft. As soon as one is detected it is interrogated by an IFF system. If the aircraft fails to give the correct coded response, the launcher is automatically turned towards the aircraft and the crew alerted. The operator views the target through an optical sight and fires a missile. Flares on the missile are tracked by television and the missile is automatically guided to its target. The system can be provided with a radar tracker, to be used instead of optical viewing in bad weather or at night.

The radar and the optical system track both the target and the missile, so that the computer can calculate deviations in the flight path of the missile and send signals to the missile to guide it to its target. Commands are transmitted to the missile by microwave radio signals and guide the missile by microwave radio signals and guide the missile moving control surfaces at the rear of the missile. The operator has only to track the enemy aircraft with his viewer; the missile is automatically guided down the line-of-sight to the target.

Rapier missiles, 2.24 metres long, 13 centimetres in diameter, with a span of 38 centimetres, weigh about 43 kilograms and have a range of between 0.5 and 7 kilometres. The missile reaches speeds greater than twice the speed of sound, fast enough to allow it to hit low-flying aircraft, even when they are travelling at high speeds. The missile carries a relatively small (0.5 kilogram) explosive warhead which is set to detonate inside the enemy aircraft, thus making it much more likely that any aircraft hit will be destroyed.

Surface-to-air missiles have essentially replaced heavy anti-aircraft guns in air defences. These missiles can reach great altitudes, change course in flight to outmanoeuvre hostile aircraft, and automatically home with considerable accuracy on their targets. However, some light anti-aircraft guns have very high rates of fire and, when operated with early-warning and control radars, are effective against low-flying aircraft, even when the aircraft are travelling at high speeds. Modern anti-aircraft guns use radar and optical tracking to provide great accuracy. They can, for example, fire up to 300 rounds per minute to a range of 4 kilometres or so.

A modern air defence system consists of a judicious mixture of a spectrum of different types of surface-to-air missiles, effective against aircraft flying at various altitudes and speeds, with radar-controlled light anti-aircraft artillery. Currently, these systems operate with fighter aircraft. The most effective American tactical air defences, for example, consist of a mixture of Patriot missiles, Stinger missiles, and F–15 (Eagle) and F–16 (Fighting Falcon) aircraft.

Air-to-air missiles

Modern fighter aircraft carry sophisticated air-to-air missiles to attack enemy aircraft. Interestingly, one of the most successful missiles in history is an air-to-air missile, the American Sidewinder. Well over 100,000 Sidewinders have been produced and sold to the air forces of many countries since the missile became operational in 1956. Sidewinders have been used with devastating effect by the Israelis against Syrian aircraft, and were used by the British in the Falklands War against more capable Argentinian aircraft.

The Sidewinder is a slim missile, about 3 metres long with a span of about 60 centimetres. Its launch weight is about 80 kilograms and it has a range of about 20 kilometres. The missile is infra-red guided and noted for its simplicity, having only about twenty moving parts and fewer electronic components than a transistor radio. As missiles go, Sidewinder is cheap. The latest model costs about $65,000.

Sidewinder has an interesting history. It was developed covertly, in the early 1950s, by a small US Naval team, led by an engineer, Dr Bill McLean, operating on a shoestring budget. McLean's team was the first in the world to tackle the problem of guiding missiles by passive infra-red homing, believing, correctly, that radar-guided missiles would be far more complex and, therefore, less effective.

The history of Sidewinder shows, by the way, how large bureaucracies lead to the choice of increasingly complex but less efficient technologies. History also shows that such large bureaucracies are very bad at innovation. In the USA, for example, each of the three services has its own huge bureaucracy to procure weapons. According to the *New York Times* (5 April 1985), the US Air Force's Systems Command employs 10,000 officers and 29,000 civilians. The US Navy's Material Command has 5,000 officers and 220,000 civilians. The US Army's Material Development and Readiness Command is staffed by 11,000 officers and 110,000 civilians. These huge agencies do not produce anything. They simply set specifications for civilian contractors. The problem is that the procurement officers insist that every part of a new weapon system incorporates the most sophisticated and up-to-date technologies, whether or not the final product is effective on the battlefield.

There is a large family of air-to-air missiles. Other air-to-air missiles, in addition to Sidewinder, used by, for example, the US Air Force and Navy, include Sparrow and Phoenix missiles, and the new Advanced Medium-Range Air-to-Air Missile (AMRAAM). Sparrow, which relies on radar-guidance to home on to its target, is larger than most current air-to-air missiles, being about 4 metres long with a 1-metre span; it has a range of some 50 kilometres.

Phoenix is the most sophisticated air-to-air missile in the world's arsenals. The 450-kilogram Phoenix missile has a 200-kilometre range and provides air defence over an area of about 40,000 square kilometres from sea level to the limits of altitude flown by aircraft. When Phoenix arrives within about 20 kilometres of its target it turns on its own homing radar to guide it to its target. A radar aboard the aircraft that launches the missiles can guide six Phoenixes to separate targets. The missile is, however, very expensive, costing about $1,300,000 each, compared with about $200,000 for a Sparrow missile.

AMRAAM is a new all-weather air-to-air missile with an active radar seeker, giving it a fire-and-forget capability. Other current medium-range radar-guided air-to-air missiles are normally guided to their targets by the radar systems on board the aircraft which launches them. The launching aircraft must, therefore, stay in the neighbourhood of the target until the missile arrives at it. This makes the aircraft much more vulnerable to enemy air defences than aircraft carrying fire-and-forget missiles.

An aircraft carrying several AMRAAM missiles will be able to engage several enemy aircraft on a single intercept. Each missile will be intelligent enough to isolate its own target, skipping over targets already chosen by other missiles. Weighing about 149 kilograms an AMRAAM is lighter than the 250-kilogram Sparrow which it will replace during the 1980s. The missile will, however, be relatively expensive. Each one will cost about $2 million, at least initially. AMRAAM is scheduled for deployment with the US Air Force and Navy beginning in 1986. AMRAAMs are autonomous enough and intelligent enough to be launched from remotely piloted vehicles. They will, therefore, be suitable for use on the automated battlefield.

Anti-ship missiles

We have seen that tanks and combat aircraft are increasingly vulnerable to attack by intelligent missiles. But, of all major weapon systems, large warships are the most vulnerable. New propulsion units, more efficient fuels, improved guidance systems, and better warheads have revolutionized anti-ship missiles.

Anti-ship missiles have an easier job than anti-tank and anti-aircraft missiles because the environment in which they fly is relatively uncomplicated. It is much easier to guide missiles accurately over sea than over land, and it is easier to locate and track enemy warships than hostile tanks or aircraft.

Several countries produce anti-ship missiles and many different types of anti-ship missiles are available. The use of these missiles in recent wars has brought home their effectiveness. As long ago as 1967, the Israelis were shocked when their biggest naval ship, the *Elath*, was sunk by a Soviet-supplied anti-ship missile fired by the Egyptians at a range of about twenty kilometres. But air-launched, sea-skimming missiles became famous during the Falklands War when the Argentinians used French-supplied Exocet anti-ship missiles to sink and severely damage British warships.

The best-known Exocet success in the Falklands War was the sinking of the British frigate, *HMS Sheffield*, on 4 May 1982. The French-built Exocet was launched from an Argentinian Navy Super Etendard fighter bomber, also bought from France, from a point about thirty-five kilometres from the *Sheffield*. The missile hit the ship about two metres above the water-line, set it on fire and sank it. The 4,000-ton warship, built in 1972 at a cost of $50 million, carried some of the world's best ship defences, including Sea Dart ship-to-air missiles. Yet it was sunk with a $250,000 anti-ship missile.

The Exocet missile, about 5 metres long, weighing 680 kilograms, and carrying a 160-kilogram high-explosive warhead, is powered by a two-stage rocket. It can be launched from a helicopter, a ship, or a land-based vehicle, as well as from an aircraft. Its maximum range is about seventy kilometres. Before the missile is launched, the rough co-ordinates of the target are fed into its computer. After the missile is launched, it skims the sea until it approaches the area in which the target is sailing. It then climbs a little, scans the horizon, detects the target with its own radar, and locks on to the target. The missile's radar, homing on the target, guides the missile to its target with great accuracy. The warhead penetrates the enemy ship and explodes inside it.

The Exocet missile that sank the *Sheffield* was one of six Exocets fired by the Argentinians during the Falklands War. Four of the six hit their targets. This success rate is less than the '90 per cent probability of destroying any type of surface vessel' claimed by Exocet's manufacturers, the French firm Société Nationale Industrielle Aerospatiale. But a 67 per cent success rate is still very good, remembering that the Argentinian forces had only recently acquired Exocets and were not familiar with them. In any case, such a success rate is very cost-effective given that an Exocet costs a tiny fraction of the cost of even a relatively small frigate.

Anti-ship missiles are of two basic types: one fired by one ship against another (ship-to-ship missiles), the other fired by an aircraft against an enemy warship (air-to-ship missiles). Typical modern ship-to-ship missiles are the American Harpoon and Tomahawk missiles. Both missiles can be fired from the standard torpedo tubes in American submarines, and in those of many other nations.

In its standard version, Harpoon is about 4.6 metres long, 34 centimetres in diameter and weighs about 670 kilograms. It has a high-explosive warhead of about 225 kilograms, of the blast penetrating type. Harpoon has a maximum range of some 100 kilometres, well over the horizon. Although normally carried on ships or submarines, it can also be dropped from aircraft.

When launched from a submarine, a Harpoon missile is contained in a capsule which has a diameter of 53 centimetres so that it just fits a standard torpedo tube. The capsule rises to the surface of the sea, and when it breaks the surface the capsule opens to release the missile and then falls back into the water, to sink without trace. Before the missile is fired, the rough location of the target is fed into an inertial guidance system on board the missile. This computer-controlled guidance system is remarkably effective – it steers the missile towards the target even if it is fired in the wrong direction. The locations of targets at the extreme of the missile's range can be determined from the parent ship's over-the-horizon radar. Harpoon carries a radar altimeter that keeps the missile flying at a relatively low altitude for most of its journey. But when the missile approaches the target area it descends until it is just above the surface of the sea. An enemy warship has a very hard job to spot a sea-skimming Harpoon coming at it.

Harpoon attacks its target in one of two ways. When it gets close to the target ship, a radar seeker searches the area, finds the target and locks on to it. The seeker commands the missile to gain height to outmanoeuvre the target-ship's defences and then dive down on the ship from above. Alternatively, the missile continues skimming the surface of the sea and strikes the target just above the water line.

At the start of its journey, Harpoon is propelled by a solid-fuel boost motor which accelerates it to a speed of about three-quarters that of sound in less then three seconds. A turbojet engine then takes over to drive the missile to its target. Once fired, Harpoon is independent of its parent ship; it is a genuine fire-and-forget missile.

Harpoon is a cruise missile. So is the Tomahawk, which can also be fired from surface ships to submarines. The missile is about 6 metres long and weighs about 1,200 kilograms. Tomahawk carries a 475-kilogram conventional armour-piercing warhead, but, like Harpoon, it can be fitted with a nuclear warhead. It is initially propelled by a solid-boost motor but uses a turbofan engine for cruising. The missile can travel long distances – it has a maximum range of about 500 kilometres – and travels at a speed of about 800 kilometres an hour.

Tomahawk can be fired from submerged submarines, is guided by a system similar to that used by Harpoon, and also flies at a relatively low altitude (10 metres or so) above the surface of the sea.

A very modern fire-and-forget anti-ship missile is the British Sea Eagle. Sea Eagle is mainly carried on aircraft or helicopters, but there is a ship-launched version. The missile weighs about 550 kilograms, is just over 4 metres long, has a 40-centimetre diameter and a 120-centimetre span. The missile is driven by a turbojet engine at a speed just under that of sound and the probable range is about 100 kilometres. The warhead probably contains about 150 kilograms of high explosive and is designed to detonate after it has penetrated an enemy ship, exploding with blast and fragmentation effects. The missile is more effective because it explodes inside the target ship rather than on impact.

Sea Eagle's sensor uses active radar. The aircraft carrying the missile tells the missile's computer the rough location of the target, and then launches the missile. The computer controls the flight path of the missile until the sensor acquires the target during the final part of the flight when the missile skims close to the surface of the sea. The sensor then accurately guides the Sea Eagle to its target.

Sea Eagle is a versatile and intelligent missile. During a test flight, for example, a Sea Eagle was successfully programmed to fly towards two ships, fly over the first and make a sea-skimming attack on the second. Sea Eagles have been shown to be capable of selecting and attacking a specific ship in a group of ships, even though the group was heavily protected with electronic countermeasures.

Intelligent anti-ship missiles like Harpoon and Sea Eagle make large warships obsolete, at least for any naval action against a sophisticated navy. Large warships are extremely expensive – a modern aircraft carrier costs about $3,000 million and a destroyer or cruiser costs about $1,000 million. Such warships can be destroyed with a high probability by, for example, a Harpoon missile costing about $800,000. In fact, the only naval ships that make military sense today are fast, small (about 200 or so tonnes) patrol boats armed with missiles and submarines. It is a sobering thought that a small missile-armed patrol boat can carry as much fire-power as a cruiser or destroyer of the Second World War.

The weapons described so far in this chapter favour the defence, in terms of cost-effectiveness. It is now much cheaper to destroy main battle tanks, long-range military aircraft, and large warships than to deploy them. This is a fact of modern military technological life.

Offensive weapon systems

Although current and foreseeable military technological developments favour the defence, the availability of new very destructive conventional warheads and the capability to deliver them over long distances with great accuracy have renewed interest in more offensive types of warfare. The emergence of these new conventional weapons has, for example, made possible proposals for changing NATO policy. Thus General Rogers, NATO's Supreme Commander, wants to use long-range missiles with very destructive conventional warheads to strike deep inside enemy territory to delay the movements of reinforcements to the front line. In this way, the General argues, he could blunt a Warsaw Pact attack with conventional weapons and thus reduce NATO's reliance on the early use of nuclear weapons.

Currently, targets at some distance from the front line inside enemy territory are assigned to NATO aircraft, such as the Tornado, a multi-role combat aircraft, and the FB–111, a bomber. But as combat aircraft like these become more expensive and their pilots cost more to train, NATO air forces are increasingly reluctant to suffer the high losses that would occur if aircraft tried to penetrate Warsaw Pact air defences. As we have seen, air defences are becoming very cost-effective. And, of course, the deeper the aircraft fly into hostile territory, the greater the risk to them. For these reasons, many military people are arguing that missiles should be used to attack targets deep inside enemy territory rather than aircraft. One such target would be a military air base. A potential warhead for attacking an air base is the British JP 233.

The JP 233 contains several hundred bomblets in one casing. When dropped over an air base, the JP 233 casing would open to release the cloud of bomblets. These are of two types. One is able to penetrate the concrete surface of runways and explode at a depth to gouge out a sizeable crater. The second type of bomblet is an anti-personnel weapon. It has a delayed-action fuse, timed to explode when people are likely to be trying to repair the damaged runway. These anti-personnel weapons explode into a large number of metal fragments and are supposed to delay the repair and make sure that the runway is out of use for a long period.

There are also proposals for a very large weapons system, called the Total Air Base Attack System (TABAS), in which a large missile would carry a 25-tonne load of runway penetrators. Just one of these weapons exploded over an air base would totally destroy it.

There are, in fact, a number of weapons for attacking large-area targets in the pipe-line. An example is the West German Container Weapon System which comes in a number of versions. Each contains forty-two tubes for ejecting submunitions. One version is a glider, a stand-off weapon with rocket propulsion and a range of about twenty kilometres. The Container Weapon System is developed from the existing West German MW–1 warhead, which dispenses about 4,000 bomblets from 224 launching tubes.

These warheads are candidates for the US Standoff Tactical Missile, now under development. Exactly what this missile will be has yet to be decided. It may be a version of one of the ground-to-ground missiles now deployed by NATO. One possibility is that a new version of the Lance missile, called the Improved Lance, will be developed, with a longer range than the model now deployed.

The current Lance missile, 6.2 metres long, with a 56-centimetre diameter and weighing 1,700 kilograms, has a maximum range of about 120 kilometres. The new Standoff Tactical Missile is likely to have a range two or three times greater than this. The range of the Lance missile could be extended considerably by, for example, providing it with a bigger motor or an additional booster. The Lance missile is carried on two amphibious tracked vehicles. One carries the erector-launcher for the missile, the other carries two spare missiles and a hoist to load them into the launcher. A lighter launcher is available for transport by helicopter or for dropping by parachute from fixed-wing aircraft.

A variety of warheads, both conventional and nuclear, have been developed for Lance missiles. One conventional type carries a cluster of fragmentation munitions; another carries a number of sub-munitions, each fitted with infra-red sensors for terminal guidance. But Lance is best known because it was suggested that Lance missiles in NATO European countries should be fitted with the notorious 'neutron' bomb, a suggestion firmly rejected by public opinion in Europe. Lance missiles (some of which are deployed outside NATO countries, like, for example, Israel) could now be fitted with conventional warheads with terminal guidance, capable of hitting targets with great accuracy.

Another type of ground-to-ground missile deployed by NATO is the Pershing. Currently, the 700-kilometre range Pershing–I is being replaced by the 1,800-kilometre range Pershing–II.

Pershing–II has a sophisticated new guidance system called RADAG. When the warhead approaches its target a video radar scans the target area and the image the radar 'sees' is compared with a reference image stored in the warhead's computer before the missile was launched. This computer controls aerodynamic vanes which guide the warhead onto its target with an accuracy unprecedented in a missile with an 1,800-kilometre range.

The accuracy of a warhead is normally measured by its Circular Error Probability (CEP), defined as the radius of the circle centred on the target within which half of a large number of warheads fired at the target will fall. In other words, the CEP is the radius of the circle centred on the target in which there is a 50 per cent chance that the warhead will fall. The CEP of the Pershing–II is a mere forty metres. And a Pershing–II missile does not take long to reach its target. Like all ballistic missiles, it travels very fast. It can, in fact, cover its maximum range in less than ten minutes. A Pershing–II missile could easily carry a conventional warhead weighing, say, 450 kilograms. And with its great accuracy, a warhead of this size could destroy large hardened military targets with no difficulty.

Conclusions

Perhaps the most far-reaching recent development in missile technology is the use of microprocessors in the guidance systems of missiles. Consequently, warheads can now be delivered onto targets with very high accuracy over long ranges. Given this accuracy, conventional warheads of moderate destructive power can effectively destroy relatively large military targets.

Another crucial development is the terminally-guided sub-munition. Warheads carrying these sub-munitions are potentially extremely effective anti-tank weapons.

Microelectronics are also steadily increasing the intelligence of missiles. Faced with accurate, intelligent missiles, battle tanks, long-range combat aircraft, and warships are, or will soon become, obsolete as manned weapons of war.

Accurate, autonomous missiles are the weapons of the automated battlefield.

Chapter 4
Reacting to Precision-Guided Missiles

The vulnerability of main battle tanks, combat aircraft, and warships to precision-guided intelligent missiles was discussed in the last chapter. Measures taken to reduce this vulnerability will be described in this chapter.

The first dramatic demonstration of the new vulnerability of tanks was the 1973 Middle East war, when more than 1,500 Arab and Israeli tanks were destroyed in a few days by anti-tank missiles and guns. The experience spurred the search for countermeasures against intelligent missiles. In the measure versus countermeasure race, the tank is losing.

So are the long-range combat manned aircraft, and the warship. In the Falklands War, for example, 114 aircraft and 10 warships were destroyed, the majority by smart, precision-guided missiles. More recently, air-to-air missiles were used by the Israelis to shoot down nearly 90 Syrian aircraft; missiles have been used to attack dozens of ships in the Persian Gulf during the Iran–Iraq war; and so on.

The tank fights back

The pro-tank lobby is, however, strong, and the world's major armies are still buying main battle tanks in large numbers. The USA, for example, wants to deploy about 7,500 M–1 tanks by the early 1990s, producing them at a rate of 70 a month. Currently (January 1985), the US Army has 2,400 M–1s in operation. The USSR is believed to be producing main battle tanks at the rate of about 220 a month. The two main military alliances already have tens of thousands of tanks in active service – some 20,000 on the NATO side, and about 50,000 on the Warsaw Pact side. And there are some 20,000 main battle tanks deployed in the Middle East.

Today's tanks are very different from their counterparts of the Second World War. They have, for example, much longer ranges. The American M–1 Abrams main battle tank, which began service with the US Army in 1980, can cover a distance of about 450 kilometres on roads, compared with about 100 kilometres for a typical main battle tank of the Second World War. The West German main battle tank, the Leopard–2, first deployed in 1978, can go even further, having a maximum road range of 550 kilometres. The Soviet T–72, introduced into the Soviet Army in 1972, has a maximum road range betweeen the other two, of 480 kilometres.

Modern tanks travel fast. Both the American M–1 and the West German Leopard–2 have maximum speeds of 72 kilometres an hour. The British Challenger, the other of the world's top-rank main battle tanks, has a somewhat slower maximum speed of about 56 kilometres an hour, similar to that of the Soviet T–72. In the Second World War, a tank speed on roads of 35 kilometres an hour was considered fast.

As anti-tank munitions have been provided with more effective warheads, the armour of tanks has, of necessity, been improved. But better armour means, in general, that the tank is heavier. Thus, the combat weights of the M–1 and the Leopard–2 are about 55 tonnes; the Challenger (which came into service in 1983) weighs even more, about 60 tonnes. These weights are 25 or more per cent higher than their Second World War counterparts. The weight of the Soviet T–72 tank is less impressive – it weighs 40 tonnes.

The increased range, faster speeds, and heavier armour of modern tanks are possible because military engineers have designed and built very efficient tank engines. The horse-power of the engine of both the M–1 and the Leopard–2 is 1,500; that of the Challenger is 1,200. These are double the horse-power of the typical main battle tank of the Second World War. Soviet tank engines, however, are less powerful than their Western counterparts. The horse-power of the T–72, for example, is 480. But most modern tank engines are powerful indeed, able to propel 55-tonne monsters at a speed of over 70 kilometres an hour. Large power-to-weight ratios are required to achieve this – about 30 for the M–1 and Leopard–2, and about 20 for the Challenger.

The most modern armour for main battle tanks is usually called Chobham-type armour; it was originally developed by the British. Chobham armour is laminated. A mixture of aluminium, ceramics, steel, and fabric, among other materials, is used. The armour is built up in layers to be more able to absorb the large amounts of kinetic energy delivered by anti-tank munitions, and the heat produced when the warheads of modern anti-tank munitions explode.

Chobham armour is thickly applied to some parts of modern tanks – the front of the chassis, the front of the turret, and the sides (the parts considered the most vulnerable). The front of the American M–1, for example, is protected by a shield of armour as much as half a metre thick. Other parts of the tank are, however, much less well protected. The top and sides of the turret, the hatches, and the rear portions carry relatively light armour. These areas are lightly protected of necessity. The tank's turret, for example, carries a heavy gun which must be rapidly rotated. This severely limits the weight of the turret and, therefore, the thickness of armour the turret can carry. And the hatches have to be light enough to be quickly opened and closed.

The upshot is that the tank is very vulnerable to attacks from the sides and the rear, and from the air. During a battle, when the tank's hatches are closed, it is particularly vulnerable to enemy infantry, armed with anti-tank missiles, attacking from the sides or rear. Even the front of the tank, thick though the armour is there, is vulnerable. The latest anti-tank warheads are capable of penetrating more than a metre of the best armour.

The most recent innovation to reduce the effectiveness of anti-tank missile warheads, in an attempt to increase the lifetime of main battle tanks, is so-called 'active armour'. Bricks containing an explosive are fixed over the armour at the front of the tank. Sensors in the bricks detect an approaching anti-tank missile and explode one or more appropriate bricks to destroy the warhead of the missile before it can significantly damage the tank. Active armour has been deployed on, for example, Israeli tanks and has been reported on Warsaw Pact tanks in East Germany.

One way of overcoming active armour is to fit anti-tank missiles with two warheads, timed to explode with a short delay. The first warhead explodes and sets off the explosives in the active armour, deactivating it, so that the explosion of the second warhead can penetrate the tank's armour. Nevertheless, the development of active armour has somewhat decreased confidence in the effectiveness of anti-tank warheads using shaped charges (see page 73) that are aimed at the front of tanks, and has considerably increased interest in anti-tank weapons that attack the turrets of tanks from the top, in anti-tank mines and in high-velocity and anti-tank cannon.

There is, of course, a relationship between protection by armour and mobility. Mobility is itself protection. The faster and more manoeuvrable the tank, the better is it able to avoid attack. But more armour means more weight and, for a given engine horse-power, less speed and manoeuvrability. Heavy tanks are, of course, more likely than lighter ones to get bogged down in mud, to do more damage to roads in peacetime (and, therefore, to have their movements more restricted in peacetime which means less training for tank crews), and to be more difficult to transport, particularly by air.

Another way to make a tank less vulnerable is to make it smaller, or, as the experts say, to lower its 'silhouette'. The sleeker the tank, the less visible it will be to operators of anti-tank missiles and guns; and, if the operators do spot it, it will be a smaller target, and therefore harder to hit. Although modern tanks are more powerful and heavier than earlier models, they are sleeker. The American M–1, the British Challenger, and the West German Leopard–2 each have heights of about 2.8 metres; the Soviet T–72 is sleeker still, standing 2.4 metres tall.

Modern tanks carry large guns. The American M–1, for example, will soon be fitted with a 120-millimetre gun, made in West Germany. Target acquisition and fire-control equipment will also be improved in another version of the tank, to be called M–1A–1. The weight of this version, to be operational in 1986, is likely to be 58 tons, three tons heavier than the M–1. Currently, the M–1 carries a 105-millimetre gun. But with a laser, rather than optical, range-finder and computerized fire control, the M–1 has greater fire-power than its predecessors. The gunner in the M–1 has an infra-red Thermal Imaging System and can, therefore, operate at night.

As American weapons expert Paul Walker, talking about the American M–1, points out:

> No longer does a tank have to stop, listen, and only then fire. Now the tank can remain hidden in defillade, accelerate quickly and advance at a high rate of speed, its gun automatically following the aim of the gunner and firing accurately on the move with new anti-armor rounds of ammunition, generate smoke for camouflage when necessary, maneuver quickly to confuse the enemy, and, when necessary, absorb or reflect many anti-tank rounds.

These words apply equally well to the British Challenger, the German Leopard–2, and, to a lesser extent, the Soviet T–72.

But, as Walker goes on to say, these sophisticated tanks have serious disadvantages. One is their thirst for fuel. In a battle, an American M–1 tank would require ten litres of diesel fuel *per kilometre*. This means, as Walker explains, 'for a company of ten M–1 tanks, at least 10,000 gallons of diesel fuel must accompany them into battle to provide each tank with more than 200 kilometres of range. Tank commanders readily admit that an easy way to stop a tank assault is to knock out the fuel supply and wait for the vehicles to run out of diesel.' And then there is the problem of maintenance. Modern tanks can travel a mean distance of only a few tens of kilometres before they break down. They generally have to be taken to the battlefield by train! Logistics is the most crucial factor in any war. For tank warfare, it is a nightmare. Quoting Walker again:

> The M–1 now requires some dozen mechanics in various specialities from chassis to turret to laser and computer areas

> to support it; units must also carry extra tracks into battle (the M–1 eats up tracks at about $25 per kilometer) as well as complete $250,000 engines and transmissions. The logistics tail for a modern armored division is therefore long, complex, and highly vulnerable to disruption. The teeth of the division – tanks, armored fighting vehicles, and other weaponry – appears increasingly sensitive to disruption in its support tail.

Armoured fighting vehicles, one of the 'teeth' of an armoured division, normally accompany tanks into battle, carrying mobile infantry, called 'mechanized infantry'. The American Bradley Fighting Vehicle, for example, provides mechanized infantry with the same mobility as that afforded by the M–1 tank. The Bradley is armed with TOW anti-tank missiles to complement the fire-power of the M–1s. The US Army intends to deploy nearly 7,000 Bradleys by the early 1990s, to go along with the nearly 7,500 M–1s to be fielded at that time.

Modern main battle tanks and their armoured companions are expensive. The US Army will probably have to spend over $33,000 million on its 7,500 M–1 tanks and 7,000 Bradleys. And what will it get for this huge sum of money? More sophistication, certainly. But not a cost-effective weapon. Again the conclusions of Paul Walker, arrived at after much research, are worth quoting:

> Armored warfare is undergoing several major technological advances, each accompanied with important potential disadvantages: improved power-plants and drive-train assemblies providing higher speed and agility but also higher cost, logistics, and maintenance; improved ammunition and fire-control systems providing better kill probabilities on the first and second shot but also complicating training and personnel support and making the turret more complex and potentially susceptible to breakdown; and lower silhouettes, less homogeneous armor, and much improved internal protection for the crew and ammunition, thereby affording greater survivability. Yet, more sophisticated armor has tended to increase weight of tanks and may not in fact guarantee any

more survivability against the most capable anti-tank homing weapons.

Anti-tank warheads

The most capable anti-tank weapons mentioned by Walker carry warheads that can easily penetrate up to a metre of armour. There are a number of different types of warhead, but the most commonly used today is either the HEAT (High-Energy Anti-Tank) round, or the Armour-Piercing Discarding Sabot (APDS) round.

A HEAT shaped-charge warhead uses a cone-shaped metal liner inside the warhead to focus its energy on the target tank. A detonator on the nose of the warhead detonates a high-explosive charge at a predetermined distance from the tank. The explosion produces a concentrated jet of molten metal which travels at about 9,000 metres a second. A metal plug is also shot out of the warhead and travels with the jet towards the enemy tank. Both jet and plug penetrate the armour of the tank. what happens then is that a stream of molten metal and very hot vapour enters the tank through the hole and rushes to fill the space inside the tank. This may kill, or disable, the crew directly or set the tank on fire or explode the ammunition being carried in the tank.

The Armour-Piercing Discarding Sabot (APDS) round, the other most common type of anti-tank warhead, is interesting because it uses uranium to penetrate the armour of tanks. The warhead unwraps as it flies and petal-like pieces fall off to leave a high-density small core – an inner 'bullet', so to speak – made of uranium, to fly faster and penetrate the armour of the target. Like the HEAT round, the Sabot warhead can penetrate up to a metre of armour on a tank and still have sufficient energy left to kill the crew or damage enough essential equipment to destroy the tank.

HEAT and Sabot rounds are much better able to penetrate thick armour than the more traditional solid shot, or kinetic energy round, which relies on mass and velocity to force its way through the target's armour. A kinetic energy round normally requires a large-calibre gun to fire it and cannot be used as the warhead for an anti-tank missile. Interest in kinetic energy anti-tank munitions is reviving, however, in, for example, the hypervelocity missile in which a relatively small warhead relies on sheer speed to ram it through the armour of tanks.

The tank is obsolete

The tank enthusiast's old adage that 'the best anti-tank weapon is another tank' is no longer true. Just as the tank made obsolete the cavalry horse, modern anti-tank missiles have made obsolete the tank, in spite of measures taken to make the tank less vulnerable. The plain fact is that it is virtually impossible to hide some sixty tons of hot metal on the modern battlefield from the sensors of intelligent missiles.

Perhaps the most critical factor in judging the usefulness of the main battle tank versus the anti-tank missile is the range at which they can engage targets. The best tank guns are not very effective beyond about two kilometres; even small anti-tank missiles, like TOW, are effective at longer ranges.

A modern main battle tank may be sleeker than older tanks and more difficult to spot by eye. But its exhaust gases typically have temperatures of over 800 degrees Centigrade. The large amount of heat given off by a tank makes it very 'visible' to sensors sensitive to infra-red radiation. Missiles guided by lasers and millimetre waves are even more effective than missiles guided by infra-red. Countermeasures available to tanks – flares, decoys, electrtonic jamming, and so on – are proving much less effective against the more sophisticated sensors. Countermeasure technology is likely to lag behind new missile-sensor technology for the foreseeable future; in fact, we can expect the relative cost-effectiveness of anti-tank warfare, compared with that of tank warfare, to continue to increase for the time being.

The tank was undoubtedly one of the most crucial developments in the history of conventional warfare. But equally certainly the tank has had its day.

Can the warship be defended?

Warships are threatened by a variety of intelligent anti-ship missiles, designed for launch from aircraft, surface ships, and submarines. Today's anti-ship missiles often travel at supersonic speeds. They also fly along paths, typically skimming the surface of the sea, that make it very difficult for threatened ships to intercept them. Anti-ship missiles are so effective, in fact, that Caspar W. Weinberger, American Secretary of Defense, recently said that anti-ship 'missiles can best be countered by detecting and engaging the ships or aircraft carrying them before they reach launch position'. But this approach is, to say the least, not a hopeful one.

One of many serious threats to American warships is the long-range supersonic Soviet AS–4 anti-ship missile. The AS–4 missile, about 11 metres long, with a span of about 2.5 metres, and weighing some 7,000 kilograms, has a range of about 300 kilometres. Backfire maritime bombers carry these missiles. Backfires, with a combat radius of some 5,000 kilometres, threaten American warships over huge areas of the world's oceans. Even if the warships escape air attack they are exposed to as great, or greater, threats from Soviet submarines. And if they escape air and submarine attacks, they can be attacked by anti-ship missiles, including cruise missiles, fired from Soviet surface ships. There is, of course, a similar spectrum of American threats to Soviet warships.

The missile threat

The US Navy, for example, is spending a great deal of money in an effort to develop a credible defence for its warships against attack by anti-ship missiles. The planned defence system recognizes that anti-ship missiles are so effective that they can best be countered by detecting and attacking the enemy ships, submarines, or aircraft carrying them before these launching platforms reach their launch positions. The US Navy's anti-air warfare programme, for example, is designed to intercept enemy bombers in an 'outer zone', before US ships come within range of the anti-ship missiles these bombers are carrying. But, as we have seen, the outer defence zone must cover huge areas of the world's oceans – so huge, in fact, that it would be impossible to monitor it fully.

In an attack, therefore, many anti-ship missiles are likely to get through. The US Navy is trying to develop defences, called 'area' defences, to attack the incoming missiles themselves at long range. Enemy anti-ship missiles that get through the 'outer zone' defences and the 'area' defences will, it is planned, be attacked by so-called 'point' defences at relatively short ranges. US warships are, therefore, to be defended by a complicated three-layer defence system in an effort to prolong the life of large warships.

Basically, the US Navy's 'outer zone' protection for its warships at sea is provided by airborne early-warning aircraft, fighter interceptors, and electronic warfare aircraft. 'Area' defences, the warships' second layer of protection, consist of long-range ship-to-air missiles carried on anti-air warfare ships, such as the CG–47 cruisers and DDG–51 destroyers now being built. The missiles carried on anti-air warfare ships protect the ships that carry them, but also neighbouring ships in a naval battle group. 'Point' defences, the third layer of defence, consist of short-range ship-to-air interceptor missiles, anti-aircraft guns, decoys, and electronic-warfare systems. They are mainly designed to protect the ships that carry them.

To try to make naval defences more effective, US warships normally sail the oceans in battle groups, headed by an aircraft carrier which today would usually be powered by a nuclear reactor. A carrier battle group includes, in additon to the carrier, destroyers, cruisers, attack submarines, and logistical support ships. Soon, some of the cruisers and destroyers will be anti-air warfare ships whose task will be to defend the battle group against enemy missile attack.

The aircraft carried on a US naval aircraft carrier include the F–14 (Tomcat), an all-weather fighter designed to protect the carrier battle group against attacks by Soviet bombers and cruise missiles. Each F–14 carries six Phoenix long-range air-to-air missiles and can launch the missiles, virtually simultaneously, against six enemy bombers. The F–14 is a very expensive aircraft; each costs the US Navy at least $40 million.

Carrier-borne F–14s are assisted by so-called wide-area surveillance, to give warning of an attack early enough to enable the fighter aircraft to get to the 'outer' zone in time to attack enemy bombers. The early-warning system will be based on over-the horizon radars, with a detection range of several hundred kilometres, operating in locations from which they can monitor the routes reckoned to be the most likely ones along which enemy bombers would approach American warships. The surveillance system, which is planned for deployment in 1987, will also be mobile, relocatable to prepared sites around the world, to establish surveillance, when considered necessary, in areas that are not routinely covered.

Early warning of attack, it is planned, will also be provided by carrier-based aircraft, specifically E–2C (Hawkeye) aircraft, equipped with look-down radars and capabilities for commanding and controlling naval air forces.

Naval 'area' defences will rely mainly on CG–47 guided-missile cruisers and DDG–51 guided-missile destroyers. Both are equipped with the complex Aegis system that uses the most sophisticated technologies to detect and intercept high-speed cruise missiles at sea. Aegis uses phased array radars to detect enemy missiles at long distances and automatic fire-control systems to track and engage many targets simultaneously. It also uses the most advanced counter-countermeasures to try to frustrate enemy jamming operations.

The anti-air warfare warships will carry ship-to-air Standard missiles. In the version having the longest range, Standard missiles, which are about 8 metres long, 34 centimetres in diameter, 90 centimetres in span, and 1,100 kilograms in weight, have a range of about 120 kilometres. The US Navy plans to buy 11,000 Standard (SM–2) missiles over the next five years, at a cost of about $600,000 per missile.

Currently the US Navy operates thirteen deployable aircraft carriers and plans to increase this number to fifteen. It believes that it needs 100 anti-air warfare ships to escort its carrier battle groups. The US Navy will have 69 of these ships by the end of

1987. These should include 22 CG–47 guided-missile cruisers and 3 DDG–51 guided-missile destroyers. *Each* of these 25 new ships will cost over $1,000 million.

A short-range ship-to-air missile intended for the 'point' defence of warships is the British Seawolf anti-missile missile. The Seawolf, about 2 metres long, 18 centimetres in diameter, 56 centimetres in span, and weighing 82 kilograms, operates with radars which detect anti-ship missiles heading for the ship. The radars feed into computers the range, bearing, and velocity of the incoming missile. When the tracker-computer calculates that the interception point is within range it automatically fires a Seawolf missile. The system can fire up to three Seawolfs in rapid succession and steer all three simultaneously to the same attacking anti-ship missile. Targets are detected and tracked, and missiles are fired, completely automatically. The whole sequence of operations is very rapidly done, as indeed it must be because the target may be a high-speed anti-ship missile. Because of the need for a very speedy engagement, a Seawolf must be accelerated very rapidly; it reaches a speed of about twice that of sound in about two seconds. The missile's range is about 6.5 kilometres.

In addition to point-defence missiles, warships may carry point-defence guns, like the US Phalanx or the Dutch Goalkeeper anti-missile systems. These radar-controlled multi-barrel Gatling guns have very high rates of fire and are designed to destroy an anti-ship missile through the detonation of its warhead by the impact of a heavy bullet. The guns have a relatively short range and the impact would take place relatively close to the ship, at less than a kilometre. Only the complete destruction of the missile's warhead would prevent the ship being hit by the missile.

The Phalanx, which weighs about six tons, has six rotating barrels. The gun fires bullets at the very fast rate of 50 per second. (The Goalkeeper has an even faster rate of fire, of 70 rounds a second.) The idea is to erect a 'wall' of bullets. An attacking missile flies into this wall of metal and is promptly destroyed.

The Phalanx gun is fully automatic and operates in all weathers. Automatic operation is essential for point-defence guns. If the gun is to hit a missile within the last kilometre of its flight, as it is designed to do, it has a mere second to hit a missile flying at about four times the speed of sound, a typical speed for an anti-ship missile. Phalanx's radar simultaneously tracks the bullets and the target, directing the bullets to the target. The bullets have a uranium core. Because uranium is a very heavy metal the bullets are highly penetrating. But the gun's magazine holds only about 1,000 rounds, enough for about a twenty-second burst of fire. The system is, therefore, easily overwhelmed if, for example, the ship is attacked by several missiles. Also, a Phalanx gun is completely worn out after fifty minutes of operation.

The submarine threat

Anti-ship missiles are not the only deadly threat to warships; another is the submarine. Once again, modern submarines are so effective if they get within range of enemy warships that by far the best way of neutralizing them is to engage them before they get within range.

To attack enemy submarines at long range, the US Navy, for example, relies mainly on its own attack submarines and long-range P–3 patrol aircraft supported by undersea surveillance systems. But a significant fraction of the submarines attacking, say, an American carrier battle group will evade the long-range detection systems. The carrier group will, therefore, use formations of surface ships carrying passive and active sonar systems and torpedo-armed helicopters for short-range protection. Anti-submarine warfare (ASW) has become an exceedingly complicated and expensive operation.

In ASW, detection is the critical element. Attempts are being made to improve detection methods by increasing the sensitivity of detectors, improving the integration between various sensing systems and improving the computer processing of data from sensors. The main categories of ASW sensors are electronic, based on radar, infra-red, or lasers; optical; acoustic, including active and passive sonar; and magnetic, in which the magnetic disturbance caused by the presence of a submarine is measured. Sensors may be carried on aircraft and ships, deployed on satellites in space, or placed on the bottom of the ocean.

The ASW activities of a superpower like the USA or the USSR are worldwide and continuous, involving a total system of great complexity including a network of foreign bases. In American ASW activities, fixed undersea surveillance systems, based on arrays of hydrophones and monitoring a large area of ocean, have long played a key role. Two new systems, one mobile and the other an air-dropped system, are being developed to supplement the fixed sea-bottom sensors.

The US Navy has also begun to deploy special ships with long-range surveillance capabilities to extend ASW coverage to those parts of the world's oceans not presently monitored by fixed, ocean-bottom systems. These specialized ships, that cost about $60 million each, are called Tagos-class Surveillance Towed-Array Sensor System (SURTASS) ships. P–3 long-range maritime patrol aircraft, which operate from a number of airfields scattered throughout the world, are provided with information from the large-area surveillance systems about the general location of Soviet submarines. The aircraft then use large numbers of sono-buoys, that detect sounds emitted by submarines, and sophisticated data-processing systems to pinpoint the submarines. ASW systems are also carried by surface ships. American ships, for example, carry the Tactical Towed-Array Sonar (TACTAS) system in which a network of sonar-buoys is towed behind a ship to detect any submarine in the vicinity. ASW helicopters, called LAMPS (Light Airborne Multipurpose System), are carried on the ships and used to attack any submarines detected – for example, with torpedoes or depth charges.

But the most effective single weapon system for detecting and attacking enemy submarines is another submarine, the hunter-killer submarine. The hunter-killer is usually a nuclear-powered submarine equipped with sonar and other ASW sensors, underwater communications systems and a computer to analyse date from the sensors and to fire ASW weapons. But hunter-killer submarines are very expensive. The US Navy is paying about $700 million for each one. Nevertheless, both the USA and the

USSR operate a large fleet of hunter-killers. The US Navy has 96 nuclear attack submarines (in addition to 4 diesel-powered attack submarines) and the Soviet Navy has about 65 (out of a total tactical submarine force of about 370).

If detected, enemy submarines can be destroyed with torpedoes, depth charges or missiles. One ASW weapon is the US Captor, a torpedo inserted into a mine-casing. A Captor, which can be stored for a long time in deep water, has an acoustic sensor and a small computer which activates the launching mechanism when the sensor 'hears' and identifies an enemy submarine. The weapon is anchored on the ocean floor to wait, a long time if necessary, for a target, and is ideal for sealing off straits to create an ASW barrier. Captor's sensor can distinguish between surface ships and submarines and has a range of ten kilometres or so. If the autonomous weapon misses its target at the first go, it can turn round and attack again.

The Americans are now developing two new long-range ASW missiles that will be able to attack enemy submarines at distances beyond torpedo range. The first, the ASW Stand-off Weapon (ASW SOW) is a new version of the currently deployed Submarine Rocket (SUBROC). The second, the Vertical Launch ASROC (VLA), is to replace the existing Anti-Submarine Rocket (ASROC). SUBROC, carried by US hunter-killer submarines, is launched from a torpedo tube. It rises to, and breaks through, the surface of the sea, flies as a missile over a distance of up to 60 kilometres, and re-enters the ocean near the enemy submarine. The missile's nuclear warhead then explodes at an appropriate depth to destroy the submarine. Other ASW missiles include the Australian Ikara, with a 20-kilometre range, the French Malafon, with a 13-kilometre range, and the Soviet SS–N–14, with a 35-kilometre range.

Existing ASW missiles are not autonomous. They rely on receiving guidance information, from the ship that launched them, during their flights, before the warhead is released. The warhead then normally homes acoustically on the sound waves emitted by the enemy submarine. Alternatively, a nuclear warhead, that is so destructive that it does not need to explode close to the submarine, is used. SUBROC's nuclear warhead, for example, is said to be able to destroy submarines within a distance of about six kilometres from the point of explosion.

Why large navies still exist

In summary, large warships are the most vulnerable of all major weapons. In spite of the enormous resources that the superpowers are putting into the development of naval anti-air warfare and anti-submarine warfare systems, large warships are becoming increasingly vulnerable to anti-ship missile and submarine attack. Warships are also much more expensive than the weapon systems that can destroy them.

Nevertheless, both superpowers still operate huge navies. The USA currently has 540 deployable warships – including 13 aircraft carriers, 2 battleships, 69 destroyers, 30 cruisers, 110 frigates, 100 attack submarines, and 37 strategic nuclear submarines. The USSR operates a navy of similar size although of a different structure, with, for example, more submarines and fewer aircraft carriers than the US fleet. The Soviets are, in fact, now building their first aircraft carrier believed to be large enough to operate modern conventional aircraft. Currently, they have three small carriers, only able to accommodate aircraft which take off and land in a short distance, and two helicopter carriers.

Given the increasing vulnerability of large warships and the escalating costs of building them (even a frigate costs about $200 million today), why are the great powers still procuring them? Probably not for any military use in war but to project power abroad in peacetime.

Traditionally, naval power has been used by great powers to influence international events. In fact, naval power has historically been used more often for political coercion than for military action. The superpowers will continue to use their large warships in peacetime to extend their influence globally. In the words of Caspar W. Weinberger, America's Secretary of Defense: 'Carrier battle groups, perhaps the most visible symbol of America's maritime capability, support our foreign policy through a series of routine overseas deployments.' Both superpowers want to play this game. In fact, the ability to project naval power globally has become a necessary characteristic of a superpower. We must, therefore, expect that Soviet–American naval rivalry in the world's oceans will increase.

Smaller countries cannot afford to buy vulnerable warships. But they are impressed with the fact that the performance and fire-power of, for example, a missile-armed modern fast patrol boat is now comparable with very much larger ships at a fraction of the cost. Hence the popularity of relatively small, but heavily armed, multi-purpose warships like, for example, the French Combattante–III, the Israeli Reshef, and the Swedish Spica-II class. Modern fast patrol boats carry all types of naval weapons – guns, torpedoes, and missiles. Automated light anti-aircraft guns and advanced ship-to-air missiles give fast patrol boats some defence against air attack.

There is also increasing interest in hydrofoils. These ships can travel at high speeds – 70 knots is typical – even in rough seas. Because their roll and pitch angles are small, the weapons on hydrofoils can be operated with high accuracy. Although today's hydrofoils are small, they can pack a hefty punch. The Italian PHM hydrofoil, for example, displaces only 60 tons but carries an automated dual-purpose heavy gun, two anti-ship missiles and an electronic fire-control system. The Soviet Saranchi-class hydrofoil is bigger, displacing some 230 tons, and carries four anti-ship missiles, two ship-to-air missiles, and a 23 millimetre six-barrelled gun. The USA plans to build large, ocean-going hydrofoils, displacing 1,000 and 2,000 tons. The People's Republic of China also appreciates the potential of hydrofoils. It operates over 120 of them.

The conclusion of all this is that although the superpowers are likely to continue to sail large warships in the world's oceans as part of superpower global politics, smaller countries will increasingly rely on smaller, cheaper ships, including diesel-powered submarines, for coastal defence, for defending extended economic zones, and to police other interests at sea.

Chapter 5
Tin Soldiers and Other Robots

'Japanese scientists said yesterday they had made a robot which could walk almost as fast as a man. A spokesman for the team at Tokyo's Waseda University said the two-footed robot called WL–1ORD took a 16-inch stride in 1.4 seconds.' (Report in the *Guardian*, 1 August 1985.)

The sons and grandsons of WL–1ORD are likely to be sold in large numbers to the military. Regiments of robots will be virtually the only fighters in future battles if military technology has its way. The battlefield will become so lethal that only the most foolhardy of humans will be persuaded to join the robots.

Consider, for example, the probable fate of a group of infantrymen in a battle in the 1990s. Their mission is to occupy an area of enemy territory. Air attacks and a fierce artillery bombardment have, they are told, virtually cleared the area of enemy troops. The infantry, therefore, spread out and advance confidently.

Almost immediately they cross the border on to the battlefield the group in the centre is totally eliminated by a fuel-air explosive. An enemy missile warhead, fired from a long way off, explodes accurately above them, producing an aerosol cloud of a substance like propylene oxide vapour. The substance when mixed with air is very explosive and the aerosol cloud, ignited when at its optimum size, produces a powerful explosion, more than five times as effective, weight for weight, as TNT. The men under the exploding cloud die from asphyxiation caused by physical damage to the membranes of their lungs. When the aerosol cloud explodes it produces a fireball that sears and burns the skin of the infantrymen on the edge of the explosion.

Soon after the fuel-air explosion the flanks of the advancing group of infantrymen, many still in a state of shock from the explosion, are attacked with fragmentation munitions. The enemy begin by firing a salvo of twelve rockets at each flank. All the rockets arrive within forty seconds. They release thousands of grenade-type munitions. When these explode they scatter small, jagged chunks of metal over a large area. The fragments have razor-sharp edges, are very hot and travel at high speed. Most of the men in range of the fragments are killed, many of them literally shredded. The few who escape immediate death are the unlucky ones. They have multiple wounds, requiring treatment from several doctors, each with a different medical speciality. There are, of course, no such sophisticated medical services near the battlefield. In any case, many of the fragmentation grenades were made of plastic. The fragments in the bodies of the survivors would not show up on X-rays. Even if doctors had been available they would have been unable to find the fragments.

The enemy continues to fire salvoes of fragmentation grenades at the attacking infantrymen with multi-launch rockets until they are all dead. None manages to penetrate more than a few hundred metres into enemy territory. The accuracy of the enemy fire is amazing. Once located, the targets are hit with the first round.

Yesterday's munitions were unguided. Once fired or released, they went to their targets under no other influence than gravity. Many bombs had to be dropped, or bullets or shells fired, to hit a fixed target, let alone a moving one; 80 or 90 per cent of munitions were normally wasted. Today, the most modern weapons are guided all the way to their targets. They are, therefore, extremely accurate even against rapidly moving targets.

If the infantry in our scenario had escaped death they would have faced other grave hazards on the battlefield. The enemy would, for example, probably sweep the battlefield with lasers to 'blind' optical sensors in the other's weapon systems. These lasers can also blind people.

An infantryman may also be attacked with new incendiary weapons that scatter very inflammable chemicals over large areas, chemicals that glue themselves firmly to the body and cause exceptionally severe burns over much of the body.

The high risk of very unpleasant death – from having the air torn out of the lungs, being shredded by fragmentation weapons, burnt with clinging incendiaries, and so on – will deter most soldiers from fighting effectively on the battlefield. Hence the military interest in the successors of robot WL–1ORD.

Functionoids

As weapons become more technologically complex, their operators must be better educated. The typical trooper is no longer the traditional, rather ill-educated 'poor bloody infantryman' but a skilled technician. Skilled technicians are so expensive to employ that no country can contemplate losing large numbers of them on the battlefield. Robots, when produced in large numbers to obtain the economies of scale, will become relatively cheap.

The manpower problem is steadily worsening for the military. In the USA, for example, the population of 18 to 25 year-olds in the year 2000 will be 20 per cent less than it is today. 'Tin soldiers' are seen as a way of solving the military recruiters' problems.

The American firm Odetics Inc., of Anaheim, California, plans to exploit the military's coming need for robots. In 1983, Odetics produced the world's first mobile, multifunction walking machine. So much is expected from this new generation of robots that a new word has been coined for them: *functionoids*. The first functionoid has been christened Odex I.

Most of today's robots are designed to perform single, repetitious jobs, mainly in manufacturing industries. Walking functionoids will perform many tasks in a variety of environments. They are in demand for work in areas too hazardous to risk human life and for exploration in places where wheeled or tracked vehicles cannot go.

Although the military use is a major spur to development, a

The British 105-mm gun is a very advanced artillery system, with a range of 17 kilometres and capable of firing a shell every 10 seconds. Carried by helicopter it can be ready for action within 90 seconds of touch-down. Very high mobility in many environments is essential for today's forces (*Royal Ordnance*)

The multi-launch rocket system is an example of a very destructive conventional weapon. Capable of firing a salvo of 12 rockets in under a minute, covering a large area with a cloud of fast, hot, razor-sharp metal fragments it is as lethal as a low-yield nuclear weapon (*Thorn EMI Electronics*)

It is not only the rich countries that develop and produce sophisticated weapons as can be seen from these examples produced by Israel's weapons industry (*Israel Military Industries*)

Today's aircraft bombs may look similar to those used in the Second World War but they are much more reliable, flexible and accurate. This Tornado is releasing a 450 kilogram (1,000 lb) bomb, with a multi-function fuse, for retarded delivery (*Thorn EMI Electronics*)

Sonobuoys detect and identify submerged submarines. The second from the right is the most powerful sonobuoy in service and provides information on the distance and direction of an enemy submarine (*Dowty Electronics*)

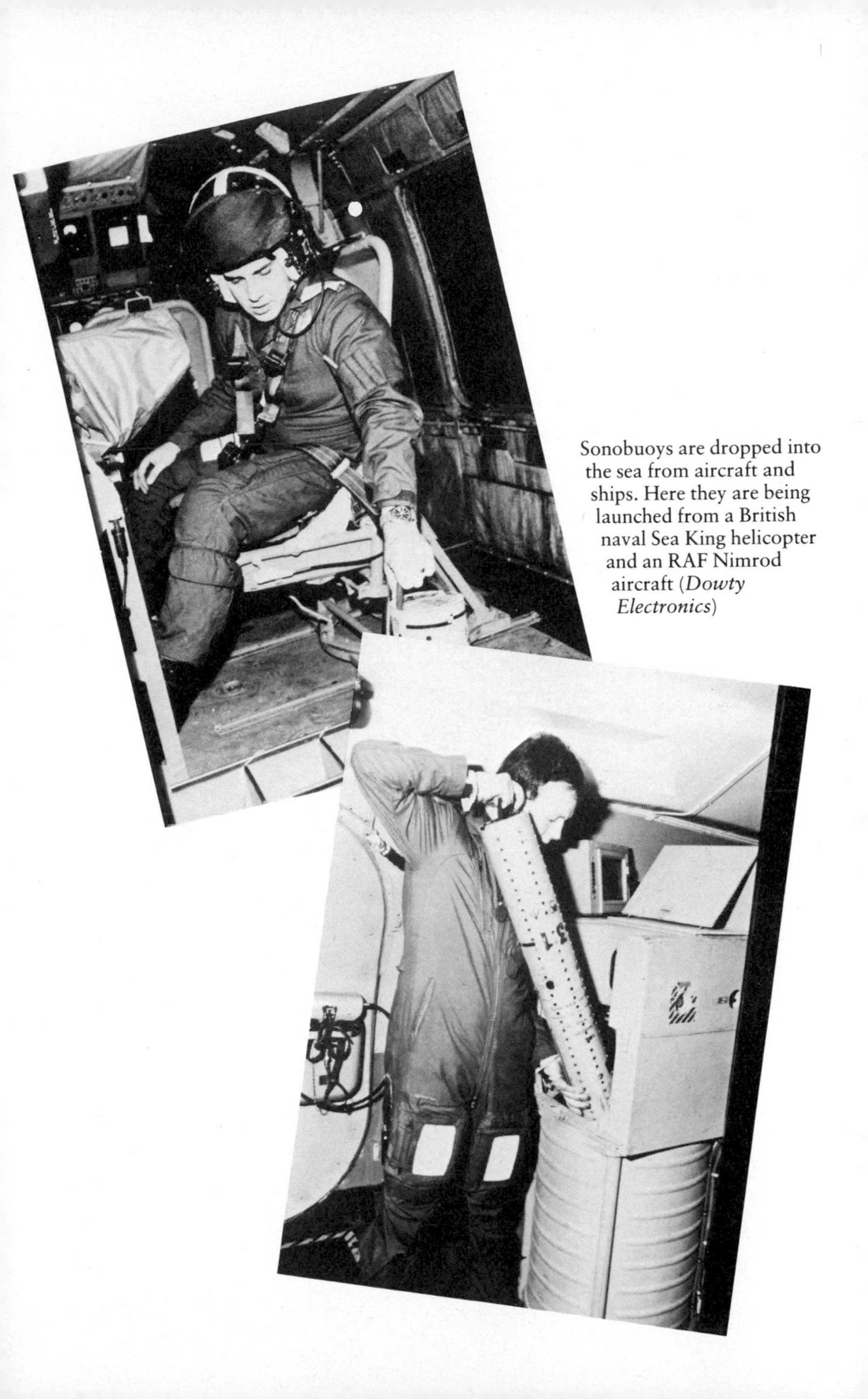

Sonobuoys are dropped into the sea from aircraft and ships. Here they are being launched from a British naval Sea King helicopter and an RAF Nimrod aircraft (*Dowty Electronics*)

Anti-tank mines can be delivered by artillery or multi-role rocket systems. This allows them to be continually 'sown' in front of advancing armour (*Dynamit Nobel*)

One of the most effective anti-tank weapons is the 'smart' anti-tank mine, like this AT-2. Scattered in front of advancing armour, these mines can destroy the leading tanks and stop the rest. The latter can then be destroyed with, for example, missiles (*Dynamit Nobel*)

Rapier is an effective British anti-aircraft missile system much used in the Falklands War in which it shot down at least fourteen Argentinian high-performance combat aircraft. Missile systems like this can be essentially fully automated. The lower picture shows a version using a laser beam (*British Aerospace*)

Light anti-aircraft guns are still effective weapons. Modern versions are extremely mobile and their operation can be virtually automated (*Marconi*)

Flycatcher is an all-weather control system for anti-aircraft guns and guided missiles used against low-flying enemy aircraft (*Hollandse Signaalapparaten B.V.*)

(*Above*) Sea Eagle is a modern air-launched sea-skimming anti-ship missile effective against warships, even those equipped with the most sophisticated air defence and electronic counter-measures (*British Aerospace*)

(*Right*) A Sea Eagle launched from a Sea Harrier. The missile system was developed and produced by the British (*British Aerospace*)

(*Above*) Photographs taken by today's military photographic reconnaissance satellites have very high resolution. This American satellite photograph shows the Soviet Navy's first nuclear-powered aircraft carrier under construction at the Nikolaiev shipyard (*Associated Press*)

(*Left*) Odex-1 is the first mobile robot, called a functionoid. It weighs 170 kilograms but can lift about 950 kilograms and can carry 450 kilograms at normal walking speed. Its grandsons will become the 'tin soldiers' of future battlefields. The US Army alone has outlined 100 applications for mobile robots (*Odetics*)

The agile six-legged Odex-1 can walk on uneven ground and step over obstacles while maintaining a stable platform. Odex-1 can assume many profiles, including 'tall', to look over obstacles, 'squat' to crawl under objects and 'tucked' for minimum exposure (*Odetics*)

vast new non-military market is expected to open up for functionoids. The nuclear-power industry wants mobile robots for reactor maintenance, inspection and emergency tasks. Mobile robots have obvious applications in mining; for example, extraction and for haulage in surface, underground and undersea mining operations. Functionoids are also being considered for oil exploration and drilling. In agriculture, mobile robots are envisioned for irrigation, harvesting, cultivating, planting, spraying and field inspection. In forestry, mobile robots could be employed for felling, planting, thinning and transporting. In the coming era of functionoids, the military will, as usual, want their fair share.

Odex I is an advance on other walking machines because it uses new electronic, mechanical engineering and computer technologies to give it impressive manoeuvrability, strength, agility, power, and speed. The six-legged robot can walk over uneven ground, climb and descend while carrying a stable platform, lift objects many times its own weight and walk while lifting. Odex I has a 'tripod gait', walking by raising three of its six articulators. These three continuously alternate with the other three, which momentarily rest on the ground to support the functionoid. Odex I can walk at a speed of about five kilometres an hour, equal to a brisk walk for a person.

The height of Odex I can be varied between one and two metres. It can carry a maximum load of about 1,000 kilograms (about six times its own weight of 170 kilograms) when stationary, with all six legs on the ground; 820 kilograms while walking slowly; and about 450 kilograms while walking at normal walking speed. Odex I has a self-contained power source, a 24-volt battery. It can walk for one hour without recharging. The functionoid receives commands from its operator through a radio link. The operator communicates with the robot's computer using a joystick control unit.

The next generation of functionoids will be autonomous, receiving only very general instructions. They will require accurate 'vision' systems, probably including stereo television with optical ranging and ultrasonics to determine dimensions and distances and perceive objects in the robots' line-of-sight. Tactile sensors will be developed for the robots' limbs. Future functionoids will not only walk but also climb, hang, swim, roll, and so on.

Functionoids will first be used by the military for reconnaissance and surveillance, transporting ammunition, sentry duty, clearing mines, and so on. As they become more intelligent they will become more autonomous. In the words of Academician V. Glushkov, director of the Kiev Institute of Cybernetics: 'At our institute, projects are under way to teach robots languages and a "rational" approach to different problems. In certain fields of knowledge, we are able to create algorithms by which a machine will learn more rapidly than even a very capable human. I will not conceal the fact that our goal is to create a "thinking" robot by the year 2000.'

Robotic land vehicles

Future functionoids will operate with autonomous robotic land vehicles, wheeled or tracked, operated by the army. These vehicles will be used as weapons platforms – firing anti-tank and anti-aircraft missiles and artillery guns; mobile mines; etc.

The development of autonomous vehicles will depend on the development of computing power. The US Defense Advanced Research Projects Agency estimates that an effective autonomous vehicle would need some 100,000 million operations a second for a single 'sense', such as vision. Today the fastest computers perform about 100 million operations a second. A thousand-fold improvement is, therefore, necessary for useful autonomous vehicles. Also the computers for autonomous vehicles must be physically small enough and must not use too much power for use in robotic vehicles. This sounds like a tall order. But given the almost incredible rate at which computers are being developed, we can expect suitable computers for autonomous vehicles to be available by the end of the century. A recent report from the Jet Propulsion Laboratory, California Institute of Technology, concludes:

It can be projected with great confidence that microelectronic

> technology will be able to produce integrated circuits containing millions and perhaps hundreds of millions of components by the year 2000. Furthermore, these circuits will be much faster than current integrated circuits by an order of magnitude or more. This will allow compact computing systems with hundreds of processors, each as fast as today's largest mainframe computers, to be located within the confines and power limitations of a conventional-size military vehicle. The total computing capacity of such a system might approach 1 trillion (million million) operations per second. This is roughly the estimated gross computational performance needed for a highly capable autonomous vehicle.

Projects to develop autonomous unmanned land vehicles are under way at the Robotics Institute at the Carnegie-Mellon University in Pittsburgh, USA, at the Artificial Intelligence Laboratory at the Massachusetts Institute of Technology, at the laboratories of the Martin Marietta Corporation in Denver, and elsewhere. The Carnegie-Mellon vehicle, for example, called the Terregator, uses lasers, acoustic sensors to judge distance, and stereo television cameras to scan the terrain ahead and compare the picture with images in the on-board computer. The information from the sensors is used to adjust speed and steering. The Terregator is a six-wheeled vehicle, weighing between 450 and 900 kilograms according to the equipment it carries, and powered by a petrol engine or an electric motor.

Remotely piloted vehicles (RPVs)

Walking robots and unmanned vehicles are now under development. But unmanned aircraft, called remotely piloted vehicles (RPVs), are already being operated by the military.

RPVs were first used as drones to test weapon systems and to train gunners and other weapon operators. They then graduated to reconnaissance work at fairly short distances. The US Air Force, for example, used Teledyne Ryan Firebee RPVs extensively for reconnaissance in the Vietnam War in the late 1960s and early 1970s. RPVs are used in a number of other armies.

The armies of West Germany, Canada, the UK, France and Italy, for example, use the CL–89 for reconnaissance. This RPV, which carries a Zeiss camera, is powered by a turbo-jet engine; it is capable of flying at 740 kilometres per hour and has a range of 120 kilometres.

These early reconnaissance RPVs carry a camera that has to be recovered and the films processed and analysed. The process takes a relatively long time. Modern RPVs, however, carry electro-optical (TV) systems. These RPVs can be used for real-time intelligence, to transmit pictures of events as they are happening, electronic surveillance, electronic warfare, artillery spotting and so on.

RPVs were most recently extensively used in combat by the Israeli Defence Force during the Lebanon war. In one operation, Israeli RPVs, carrying electronic equipment, were flown near Syrian surface-to-air missile batteries to detect the frequencies of the radars used in order to fire and control the missiles. The information served to calibrate sensors on Israeli missiles. The Israeli missiles were then used to attack the Syrian air defence system; the Israeli missiles were able to home on the Syrian radars to destroy them and the surface-to-air missiles associated with them. The Israelis also flew RPVs for target identification, artillery spotting, and electronic countermeasures.

The Israeli Defence Force uses both Scout RPVs, built by Israel Aircraft Industries, and Mastiff RPVs, built by Tadiran, a subsidiary of the United Industrial Corporation.

The Scout RPV carries a TV camera aimed and focused by a remote operator on the ground. The operators had no problem in spotting vehicles on the streets of Beirut from the comfort of their home base, or in tracking helicopters in flight.

A Scout is similar in size to a Mastiff RPV. The Mastiff Mark III, a fixed high-wing aircraft with a pusher engine installed in the rear, is 3.3 metres long and has a wingspan of 4.2 metres. It weighs 61 kilograms empty and can carry a payload of 30 kilograms; the maximum take-off weight is 115 kilograms. The Mastiff has a maximum speed of about 200 kilometres an hour when fully loaded and cruises at 115 kilometres an hour. Its operational ceiling is about 3,000 metres and its operational range is over 100 kilometres. The RPV can take off from short improvised runways or can be launched from an hydraulic launcher mounted on a truck.

The RPV and its payload can be remotely controlled from a fixed ground control station or a portable control station, or independently controlled by an airborne automatic pilot control. The use of a portable control station extends the operational range to over 200 kilometres.

The ground control station, normally carried on a military truck, contains all the command, control and communications needed to control the RPV and its payload, and to receive and process data transmitted by the RPV. The system tracks the RPV automatically. Commands are transmitted from the control station to the RPV's auto pilot and payload by radio. If the RPV carries a television camera, it transmits video signals from the television camera to the control station.

The portable control station, which consists of two boxes mounted on tripods, one to control the RPV, the other to control the payload, facilitates take-off and landing of the RPV at a site remote from the ground control station. Information from the RPV can then be transmitted to the users in or near the ground control station while the RPV takes off and lands near the portable station. The portable station can also be used to extend the range of the RPV.

The landing system is a hook and wire system in which a hook is connected to the airframe and a wire stretched above the runway between two energy absorbers. When the aircraft lands, the hook engages the wire and the absorbers dissipate the landing energy and bring the machine to a stop within about eighteen metres. A section of a highway can be set up as a landing strip in a few minutes.

The Scout and Mastiff RPVs, like other operational RPVs, are small; they are, in fact, called mini-RPVs. Much larger RPVs are under development. These will carry several hundred kilograms of TV equipment, radars, communications, and so on. They will operate at very high altitudes, of 25,000 metres or so, and will reach speeds of 500 kilometres per hour and more. RPVs are under active development in West Germany, Italy, France, the UK, and the USA.

Future RPVs will carry a variety of other sensors – infra-red, radar, millimetre-wave, and so on. They will be fitted with sophisticated antennae and other communications equipment, laser designators, and electronic equipment for use in electronic warfare, and to provide electronic countermeasures. They will also carry a range of weapons.

RPVs, with their infra-red and millimetre-wave sensors, will operate at night, in all battlefield environments and in bad weather. They will, in other words, provide a 24-hour service. They will be used for a wide variety of missions, in addition to real-time continuous intelligence, including communications links, electronic warfare operations, laser designation, target acquisition, and weapon delivery.

An example of an RPV now under development is the KZO (Kleinfluggerät für ZielOrtung) that would carry sensors for day and night surveillance and target acquisition in all weathers at ranges of 50 or so kilometres. The 2.2-metre-long RPV will cruise at 220 kilometres an hour, for more than three hours, at altitudes between 300 and 3,000 metres. The KZO, to be launched by rocket and retrieved by parachute, will carry a number of sensors including infra-red, television, and laser. It will weigh about 100 kilograms and have a payload of about 45 kilograms. Other similar examples are the US Aquilla and the British Phoenix systems.

Some of the RPVs under development fly by recognizing the pattern of the landscape. The flight-path can be pre-programmed in the RPV's computer, or a TV camera carried by the RPV can transmit a picture back to an operator behind the lines. He can then decide whether to keep the aircraft flying in a certain area over enemy territory or move it to another sector. One ground control station will control several RPVs.

Because they carry no pilot, RPVs can be made very small and can be submitted to accelerations high enough to injure a pilot in a manned aircraft. Their small size gives them a relatively small radar cross-section. This makes them difficult to detect, especially when they fly at high altitudes.

In future wars, RPVs, carrying air-to-air missiles, will be used for air-to-air combat and, carrying air-to-surface missiles, for ground-attack missions. The West German firm MBB, for example, is developing an anti-tank RPV, called the Panzer

AbwehrDrohne (PAD). PAD, which carries acoustic and millimetre-wave sensors, works in combination with, say, a KZO RPV. The KZO detects enemy tanks and guides a PAD to the area. The noise of the tank engines is detected by the acoustic sensor and causes the RPV to dive towards the tanks. The millimetre-wave sensor is then used to guide the PAD to a tank which is destroyed by a warhead carried by the RPV.

RPVs will increasingly replace manned aircraft. There is no technological reason why RPVs should not do all the jobs manned combat aircraft do.

Largely because there is no need to defend a pilot, which requires a great deal of expensive electronic and other equipment, RPVs are relatively cheap. Assuming that in an air attack across an air defence system there is an attrition rate of 20 per cent, then if 500 aircraft make three sorties, 244 aircraft would be lost. At $25 million an aircraft, this would be a loss of $6,100 million. This money would buy more than 20,000 RPVs at $300,000 each.

An idea of the economics of RPV systems can be had from MBB's estimate, quoted in the journal *International Defense Review*, of the cost of the KZO. A platoon of ten KZOs with its ground control station is likely to cost about $5 million. An expendable PAD is likely to cost about $19,000.

Artificial intelligence

Dr William J. Perry, formerly the Pentagon's Under Secretary of Defense for Research and Engineering, recently said that 'during this decade, there will be a thousandfold improvement in cost performance of computers; that is, computers available to us in 1990 will have a thousand times the cost performance of computers that were available in 1980. And a thousandfold improvement is so large that it is difficult to plan for the future.' By 'cost performance' Dr Perry meant the cost of buying the computer, plus the cost of operating, maintaining and programming it. Dr Perry, who was responsible for weapon procurement and research and development, explained that during the

1980s the geometry of microchips will go from about 5-micron spacing of the circuit elements on a chip to about 0.5-micron. (A micron is one-thousandth of a millimetre or one-millionth of a metre.) This, incidentally, is the objective of the Pentagon's ongoing Very High Speed Integrated Circuit (VHSIC) programme. The integrated circuit technology research has already produced functional 1.25-micron VHSIC devices.

Reducing the spacing of the circuit elements on a chip from 5-micron to 0.5-micron will allow 100 times as many transistors to be put on the chip. History shows that cost performance is proportional to the number of transistors per unit area of chip. Improvements in hardware will, therefore, lead to a 100 to 1 improvement in cost performance. A further improvement in cost performance will come from improvements in software and in computer design. According to Dr Perry, the introduction by the military over the next few years of a sophisticated new computer language, called Ada, and other software changes will improve cost performance by another factor of ten. Combining this with the hardware improvement of 100 to 1 will produce the overall improvement, during the 1980s, of 1,000 to 1 predicted by Dr Perry.

The rapidity with which computer performance is improving, and consequently the rapidity with which the size of computers is being reduced, are often forgotten. The pocket calculators we are so familiar with today would have filled a large room twenty years ago.

Work on computers called 'fifth-generation' computers, a thousand or more times more powerful than current computers, is going on mainly in the USA and Japan.

In the USA, in the civilian sector, there is a joint effort by some large US computer companies, and IBM is working separately. Military research is being undertaken by the Defense Advanced Research Projects Agency (DARPA). Japanese research and development on fifth-generation computers is coordinated by the Institute for New Generation Computer Technology programme. The competition between the Americans and Japanese in this field is cut-throat. DARPA's Strategic Computing programme is developing a class of 'super-intelligent' computers for application to advanced defence systems by the end of the 1980s. These new computers will be capable of 'vision' for robots and autonomous vehicle navigation, 'understanding' the English language, and 'speech recognition' for use in an aircraft cockpit. It is claimed that 'small-scale feasibility demonstrations of these concepts have been carried out in the laboratory, but they need to be engineered for application to practical defense systems'.

But another, and in the opinion of many experts an even more profound consequence of greatly increased computer power will be the introduction of artificial-intelligence technology. The prospects for artificial intelligence were discussed by British computer designer Sir Clive Sinclair, in a speech delivered on 29 March 1984 to the US Congressional Clearinghouse on the Future.

> The human brain contains, I am told, 10 thousand million cells and each of these may have a thousand connections. Such enormous numbers used to daunt us and cause us to dismiss the possibility of making a machine with human-like ability but now that we have grown used to moving forward at such a pace we can be less sure. Quite soon, in only ten or twenty years perhaps, we will be able to assemble a machine as complex as the human brain and if we can we will. It may then take us a long time to render it intelligent by loading in the right software or by altering the architecture but that too will happen. I think it certain that in decades, not centuries, machines of silicon (the basic material used in the electronics industry) will arise first to rival and then surpass their human progenitors. Once they surpass us they will be capable of their own design. In a real sense they will be reproductive.

What do we mean by 'intelligence' when talking about computers? An intelligent computer must be able to do more than calculate; it must 'think'. It must be able to accumulate knowledge and apply it to solve new problems. It should understand language. And, many would add, it must be able to make judgements. It must, in other words, be able to do things that require reasoning and perception.

How do we know if a computer can think? A famous

computer scientist, Alan Turing, believed that one could test whether or not a computer can think by putting the computer in one room, a person in another room, and an examiner in a third room. The examiner would not be told which room contained the computer and which contained the person. The examiner would communicate with the computer and the person by teletype and would ask each as many questions as he wanted. If the examiner was unable, from the answers to the questions, to say which was the computer, then, in Turing's opinion, it would be fair to say that the computer could think.

Today's computers would fail the Turing examination. But, according to experts like Dr Perry, computers at the end of the 1980s with a thousand-fold improvement in performance will be 'approaching that level of performance'; they may well graduate in the early 1990s.

One of the most active current fields in artificial intelligence is the field of 'expert systems'. Expert systems use computers to store knowledge about a particular subject and use it to solve problems. Some expert systems are already in routine use, specifically in organic chemistry, medicine and computer design. In organic chemistry, there is a routine programme for chemical analysis, specifically to analyse mass spectrographic patterns to find the structure of unknown organic chemical compounds. And there is another programme to interpret proton electrophoresis results. In medicine, a programme to interpret pulmonary function tests routinely is available. And computers are normally used in the assembly of other computers. Xcon is an expert system that automatically works out the physical layout of computers and the interconnections of their components.

Other expert systems under development, but not yet in routine use, include programmes for medical use (to treat infectious disease, to diagnose problems in internal medicine, and so on), and programmes in the fields of mineral exploration (to identify ore deposits, select drilling sites, etc.) and ocean surveillance.

From experience with expert systems we know that computers that replace some human experts are feasible, particularly in fields in which problems are solved mainly on the basis of a large body of facts and empirical knowledge, as opposed to, for example, the application of some common sense. There is, however, also experience in the use of computers in fields such as design which require the application of perception and synthesis. The potential of artificial intelligence for use in expert systems is clearly large.

There has also been progress in applying artificial intelligence to language understanding and processing. Programmes are available to answer questions asked in English, to translate sentences from one language into another, and to follow instructions given in English. Computers have been made to receive instructions by voice – instructions spoken through a microphone rather than typed into the computer. Clearly, computers will be able to communicate with humans in natural language, like English.

Another area of artificial intelligence that is progressing rapidly is pattern recognition. Programmes have been evolved that can recognize objects in visual scenes and that can identify small changes in pictures. Processing visual information is important in military applications – for example, to enable missiles to pick out and identify tanks in the presence of other types of vehicles.

Conclusions

The consequences of foreseeable increases in computer power will include the development of vastly improved real-time military intelligence and target acquisition systems; the development of intelligent autonomous robots and unmanned vehicles; and the development of very intelligent weapon systems, particularly extremely accurate missiles.

Artificial intelligence will be much used in military decision-making. In particular, command, control and communications will be increasingly handled by computer.

Chapter 6

See Cubed Eye – Command, Control, Communications and Intelligence (C3I)

Put in the simplest way, a typical battle takes place in four distinct stages. In the first stage, the enemy forces are located, identified and tracked. The aim is to find out where the enemy will attack and with what forces. In the second stage, the threat posed by the enemy forces is evaluated and decisions are made about how to deal with it. Then, in the third stage, appropriate weapons are chosen and fired at the enemy forces. In the final stage, the damage done to the enemy forces is assessed by reconnaissance to see if the enemy forces have been made ineffective. If they have not, the sequence is repeated until they have been made ineffective.

The first stage of the battle is, of course, carried out by military intelligence (I). The second and third stages are the C3 parts of C3I.

In Chapter 2 the ways in which the military collect information about enemy forces was described. As we saw, military intelligence uses a vast network of sensors – early-warning-of-attack and other reconnaissance and surveillance satellites, ground sensors, airborne and ground radars, infra-red and millimetre-wave sensors, and so on. The aim of military intelligence is to provide information about enemy forces in real time.

The first stage also involves the use of navigational equipment to determine accurately the position of one's forces and to determine precisely the co-ordinates of targets for one's weapons.

Today, so much information is produced by reconnaissance, surveillance, navigation and other sensors that it must be analysed by computer to sort out useful information from the rest. Military information gathering, therefore, of necessity goes hand-in-hand with the use of the most sophisticated computer processing in military command and control centres. Increasingly, artificial intelligence will be used by the military for data processing and decision-making. In tomorrow's battlefield, the computer will be the essential element in all C3I systems.

In advanced countries, which operate sophisticated military intelligence organizations, it is normal to place the four parts of C3I under one management that, in the words of the US Joint Chiefs of Staff, performs its task 'through an arrangement of personnel, equipment, communications, facilities, and procedures which are employed by a commander in planning, directing, co-ordinating, and controlling forces and operations'. To indicate the importance the superpowers put on C3I, the USA spent about $30,000 million on it in the fiscal year 1984, of which about two-thirds went on tactical communications, electronic warfare and intelligence.

Ann Marie Cunningham and Mariana Fitzpatrick, in their book *Future Fire*, describe C3I as a meshing of Command and Control (C2) *functions* with two sets of *systems* that allow the performance of C2 tasks. They list C2 functions as:

monitoring enemy troop strengths and resources;
monitoring one's own troop strengths and resources;
planning and replanning electronic warfare scenarios;
assessing warning signals and evaluating attack damage;
monitoring specific conflict situations:
choosing from among operational options and facilitating their execution;
assessing and controlling remaining military capabilities;
reconstituting and redirecting forces;
negotiating with the enemy and terminating conflict.

They list the systems that make C2 function as:

communications systems (the C that boosts C2 to C3) that ensure that forces and data (intelligence) sources are connected;
information-gathering and -processing systems (the I in C3I).

Of the C3I operations, the importance of communications cannot be overestimated. Although this is true of warfare in general, it is particularly true for automated warfare. The more unmanned vehicles are used, for example, the greater the need for rapid and secure communications.

Static weapons, like smart anti-tank mines, that can identify specific vehicles and explode at the appropriate moment to destroy an enemy vehicle, are completely autonomous and have no need to communicate or be communicated with. But weapon systems that move do need communications.

And, if the mobile weapon systems are fully automated, communications with them must be very frequent, if not continuous. The development of rapid communications for automated warfare that can operate effectively when the enemy is trying to jam and frustrate them is perhaps the most difficult to achieve.

Global military C3 networks

The USA and the USSR have deployed vast military communications systems that are essentially global networks. No other country can compete with these systems. The US military communications system is one of the world's most complicated and sophisticated electronic networks, second in size only to the network operated by AT&T, America's biggest telephone company.

There is, in fact, an enormous gap between the communications capabilities, and hence the military capabilities, of the superpowers on the one hand and their nearest rivals, such as France, the UK, and West Germany, on the other hand.

At the centre of America's military communications system is the US Worldwide Military Command and Control System, a complex network of information-gathering and communications systems, on land, at sea, in the air and in space. The USA spends at least $1,000 million a year on the system and employs about 90,000 people just to operate the communictions and command centres. According to Eric J. Lerner, an American defence expert, the US Worldwide Military Command and Control System uses some 35 large computers at 25 communications centres, with over 17 million lines of programming. No smaller country can compete with such a gigantic enterprise.

The purpose of the system is to enable the United States National Command Authority (the President, the Secretary of Defense, and their authorized successors in the chain of command) to have operational control of their military forces at all levels of combat. This includes providing the means by which the National Command Authority can receive information about enemy actions in order to make its decisions. Military missions are assigned to the various military commands and communicated to commanders through the global system.

At the centre of the global command and control web is the National Military Command System which includes the command centres and the communications used by the National Command Authority. The command posts include the National Military Command Center at the Pentagon; the Alternate National Military Command Center at Fort Ritchie, Maryland; the North American Aerospace Defense Command (NORAD), at Cheyenne, Colorado; and the Strategic Air Command (SAC), at Omaha, Nebraska. The National Command Authority also has an airborne command post, which would operate if there was a danger that a war would become nuclear. The Commander-in-Chief of the Strategic Air Command keeps an airborne command post continuously airborne.

In addition to this main command system, the American Commanders-in-Chief of the European, Atlantic, and Pacific areas have their own fixed and airborne command posts capable of communicating with American military forces wherever they are. Worldwide communications links are provided by communications satellites. Satellite terminals are installed at key points, including airborne and ground-based command centres, reconnaissance aircraft, and so on. Also involved is a global network of military bases including those at North West Cape, Australia, Diego Garcia in the Indian Ocean and Guam in the Pacific. There are also global networks of radars to track, control, and interrogate US military satellites.

The Defense Satellite Communications System, a super-high-frequency satellite communications system using six satellites, is the communications link between the American continent and American military forces abroad. Large fixed and mobile terminals based overseas link in with the Worldwide Military Command and Control System and tactical communications systems. The USA is also now deploying an eight-satellite system, called MILSTAR, to improve its communications with its forces worldwide. The system, jam-proofed and manoeuvrable to evade attack, should be in place by 1990.

The Defense Communications System provides American military forces with global voice, data and teleprinter services. This is being made more flexible and interoperable with the systems of America'a allies. New electronics are being used to make the main radio links more secure against jamming, interception and eavesdropping. According to Lerner, the Defense Communications System includes, in addition to 6 satellites, about 600 facilities, 100 ground satellite terminals, and some 30 million miles of wire. The system includes an automatic voice network, called Autovon, that provides the telephone service for the Pentagon; an automatic data network, called Autodin, that transmits teletype messages worldwide; and an automatic secure voice communication system, called Autosevocom, that provides encoded telephone services.

Currently, improvements in American tactical communications systems have high priority. New systems include the Ground Mobile Forces Satellite Communications system to provide reliable, jam-resistant communications support for battlefield commanders. This uses satellite communications to link military headquarters in the field to permit commanders to transmit orders and intelligence data over long distances. The plan is to spend $1,035 million on this system in the fiscal years 1984–1987 inclusive.

The Single-Channel Ground and Airborne System will provide very-high-frequency radios for battalions and companies, more jam-resistant than the radios now used by the US Army. The Army intends to deploy 45,000 of these new radios at a cost of $780 million, or $17,300 for each radio.

The Army Data Distribution System is a new digital communications system for command and control, intelligence, air defence, fire support, electronic warfare, and other military computer systems. The Manoeuver Control System is a network of small computers to provide tactical commanders with information on the location of their own forces and those of the enemy. $300 million will be spent on these two systems in the four-year period fiscal 1984 to 1987.

The Joint Tactical Information Distribution System is a new secure, jam-resistant voice radio designed to enable a military commander to put an entire battle force on the same communications network and transmit information to everyone on the network in real time. The Joint Tactical Communications programme will provide modernized switched voice and digital communications using tropospheric scatter radios. And the Tactical Air Control System Improvements programme will provide more modern command and control capabilities to control air attacks more rapidly against mobile ground targets. The Pentagon intends to spend $4,200 million on the development and procurement of these three systems between the fiscal years 1984 and 1987 inclusive.

Given the increasing demands on military communications as military operations become more automated, it is hardly surprising that the superpowers are spending huge sums on improving them. The USA, for example, is spending billions of dollars a year on computer software, often the weak link in automated military systems. Effective software for many weapon systems has yet to be developed, even though the computer hardware exists.

In the USA, for example, the Software Technology for Adaptable, Reliable Systems (STARS) Program is planned to create a system of computer-aided techniques and methods for the development and support of appropriate software. The objective is to provide a ten-fold reduction in the cost of software and a similar reduction in the rate of latent defects in software systems. This programme involves the creation of a Software Engineering Institute that will have the purpose of reducing the lag in applying new software technology to military systems.

The Pentagon is, in fact, funding some of the most advanced

work in computer research. One such project is research on very-high-speed-integrated circuits and very-large-scale-integrated circuits, aimed at producing ultra-small and very fast silicon microchips that are more reliable than the chips currently used in weapon systems.

C3I systems will benefit greatly from the use of advanced integrated circuit technology in military equipment. New technologies will be used in an effort to evolve survivable computer communications, secure message and information-transfer systems, crisis-management and command systems, and micro-computers.

The development of micro-computers will be of particular importance for improving C3I operations. We now know that computers need very little, and perhaps no, power to operate them. This means that they need produce very little heat. Hence extremely small computers will be produced.

Reduction in the size of computers and the development of appropriate software will make easier the solution of one of the main problems in C3I operations. So much information has to be handled in a modern military command centre that it is essential to sort out rapidly useful information from useless signals, called 'noise'. The increased computer power that micro-computers will make possible will be used to allow command-centre personnel to cut through noise more rapidly.

Because warfare is becoming increasingly mobile, modern armies are relying more and more on satellite communications (even individual soldiers often carry communications back-packs that allow them to communicate via satellite) and line-of-sight microwave relays. Armies are moving away from telephone communication via land line and are reducing their reliance on long-range high-frequency and very-high-frequency radio, except for communications with very mobile units. Short-range radios will be made more secure by the use of frequency-hopping techniques that automatically and regularly change the frequency of transmissions. The aim is to make communications more mobile as warfare itself becomes more mobile.

C3CM – frustrating enemy C3 systems

C3 operations are so crucial to military operations that, if the enemy's C3 can be disrupted, battles can be won before they start. Great efforts are being made, therefore, to perfect command, control and communications countermeasures, called C3CM, designed to disrupt the enemy's control of his forces and delay his decision-making process. In effect, C3CM attempt to paralyse the military body of the enemy by eliminating his military's central nervous system, his C3.

The more rigid the control of the military forces and the more centralized the C3 system, the more vulnerable it is to attack and disruption. C3 centres can be attacked with weapons, such as anti-radiation missiles, smart artillery shells, smart bombs, etc., and disrupted by so-called 'soft-kill' activities, like the use of chaff, electronic jamming, deception and so on.

C3CM is part of electronic warfare, since electronic warfare weapon systems include means of disrupting enemy C3 systems. The US Air Force's EF–111A is specially designed to suppress enemy air defences by electronic means, and therefore has some C3CM role. The aircraft carries equipment to suppress enemy long-range detection and acquisition radars, but it is also equipped for short-range jamming missions.

The EA–6B, that operates from aircraft carriers, is the US Navy's equivalent to the EF–111A. It is a sophisticated electronic warfare aircraft designed to degrade enemy defences by jamming their radars and communications.

COMPASS CALL is a US airborne jamming system, carried on EC–130 aircraft, that is specifically designed to degrade an enemy's C3 capability. AWACS aircraft, like the E–2C, operating from US aircraft carriers, and the US E–3A, are designed to operate both as C3 centres and as C3CM systems. In particular, these aircraft attempt to jam enemy communications systems.

The US Army, also anxious to increase its ability to disrupt enemy C3 systems by jamming communications, is deploying modern tactical jamming systems, EXJAM hand-emplaced expendable jammers, and EH–60 Quick Fix helicopters fitted with C3CM equipment.

Although only a few years old, co-ordinated C3I has provoked its countermeasure, C3CM, that has already become an established military activity. Future battles may well be decided by who wins the initial C3I–C3CM skirmish.

War games

Attempts to understand the way future battles will evolve are made by playing war games. War games are also used by the military for training purposes, for operational research and military planning.

War gaming is an old activity. The war games played by the German General Staff before the First World War are well known (although the results of these games, that showed the inadequacies of the German forces, were ignored to the peril of the German Army).

The activity got a bad name during the Vietnam War when the Pentagon used it to simulate fighting between US forces and the insurgents. Political and human factors were ignored and, not very surprisingly, the results were completely false. But they were used to make very over-confident predictions about the performance of American forces in Vietnam, much to the cost of those forces.

Nevertheless, war games have been re-established as a routine military activity. Using computers, very complex war games have been developed.

Andrew Wilson, former defence correspondent for the *Observer* and now its foreign editor, has made a study of war games. He defines the term 'war games' to include a wide variety of military exercises, ranging from tactical exercises conducted on the ground by skeletal headquarters staffs, to elaborate representations of a complete nuclear war played entirely by computers. In tactical war games, Wilson explains:

> the 'Blue' and 'Red' commanders are physically separated. Each has a model, generally a map, of the ground over which operations will take place. Symbols for the opposing forces

> are moved on the model according to the commanders' decisions, and the movements are duplicated on a separate model by the direction staff, or 'Control'. The outcomes of moves are determined by set rules ('rigid assessment') or by the judgement of umpires ('free assessment'). Some rules may be absolute, for example, those governing the speed of unopposed movement. Others will be 'probabilistic'; that is, they assign a certain latitude to the outcome of an event and leave it to a random number device to determine the exact result.

Wilson believes that such games, which he calls 'manual' games, are 'relatively faithful in the representation of complex situations'. They are played at NATO headquarters, the Pentagon's Joint War Games Agency and the UK's Defense Operational Analysis Establishment. The military use them for operational planning purposes.

In some 'manual' tactical games, particularly in naval and air war games, parts of the process are done by computer. In one example described by Wilson, played by the US Navy Electronic War Simulator at the Navy War College at Newport, Rhode Island, the movements of up to forty-eight warships or aircraft can be represented electronically. Electronic countermeasures, weapons' malfunction, susceptibility of units to radar or sonar detection, and so on, are also included and the battle proceeds on a huge screen.

Wilson lists a number of activities that complement war games in assisting military planning, the evolution of C3 operations and probing the weakness of the enemy's C3. These include systems analysis, the study of particular objectives in relation to the availability of resources; military case studies, the analysis of historical, as opposed to simulated, situations; general operational analysis; and game theory.

As the battlefield becomes more automated, the battle itself becomes more like a war game. As Wilson points out, war games are 'relatively faithful in the representation of complex situations. They include an element of chance, which is an essential ingredient in war. And those who design them are hopefully made aware of parts of a situation about which there is little or no data.' As they are repeated with changed parameters, they become more realistic and credible. The use of computers makes many repetitions feasible.

Where will it end? Will the battle eventually become so similar to the war game that the one becomes the other?

Chapter 7

The Militarization of Space

At a place near Colorado Springs, within sight of the Rocky Mountains, the US Air Force is building its new Consolidated Space Operations Center (CSOC). The Center, which will cost more than $1,000 million, will control American military space operations. The fact that the US Air Force Space Command needs this sophisticated new complex is a good indication of how fast space is becoming militarized.

Colorado Springs is also the home of the North American Air Defense Command, or NORAD, with its nerve centre built deep inside Cheyenne Mountain. NORAD's main task is to give early warning of a missile or air attack on North America. But NORAD, a joint American and Canadian set-up, has another job, to keep track of the more than five thousand objects in space – operating satellites, 'dead' satellites, bits of old satellites and their launchers, and sundry other space 'junk'.

The co-location of CSOC and NORAD at Colorado Springs is, of course, no coincidence. When the Space Operations Center is completed the Commander of the US Air Force Space Command, who incidentally doubles as the Commander-in-Chief of NORAD, will be able to control US military satellites and shuttle missions, and keep track of all the objects in space, from the same headquarters.

Not to be outdone, the US Navy has created its own Space Command. Although this creation is mainly related to inter-service rivalry, the US Navy is anxious to stake out its not insignificant claim on space activities.

The military use of space

Weapons in date back to the last years of the Second World War when the Germans invented the V–2 missile. Part of the trajectory of V–2s on their way to their targets in England was spent in space. The early V–2s typically reached a height of about 100 kilometres, the greatest height reached by any man-made object up to that time.

The success of the German V–2 stimulated the superpowers to develop intercontinental ballistic missiles. Whereas the range of the V–2 was about 350 kilometres, the range of intercontinental ballistic missiles is typically about 13,000 kilometres. On normal trajectories these missiles reach heights of some 1200 kilometres, well into space.

In the past twenty-five years, the USA and the USSR have deployed large numbers of strategic ballistic missiles; Britain, France, and China have also become strategic nuclear powers. And India and Japan have developed launchers to put spacecraft into orbit; these rockets could easily be used as military missiles to deliver warheads over ranges of several thousand kilometres.

Until 1981, the rockets that launched objects into space could only be used once. But we have now entered a brand new era of the space age, the era of the re-usable launcher. This began on 12 April 1981 when the first US Space Shuttle was launched.

The Space Shuttle, which normally flies in an orbit about 300 kilometres above the Earth, is designed for many tasks. It can place satellites into orbit, it can snatch satellites out of their orbits, it can repair and service satellites in orbit, and so on. It could also be used to send a satellite on its way into a very high orbit, or send a satellite into an 'escape' trajectory that would allow it to get right away from the Earth and travel on and on into space. The Space Shuttle could, of course, 'kidnap' a Soviet satellite from its orbit. It could, therefore, be seen as the ultimate anti-satellite weapon.

The Soviets are also experimenting with re-usable launchers. Delta-winged re-entry vehicles have been launched into orbits about 200 kilometres high and recovered after spending some time in orbit. Cosmos 1517, for example, was a delta-winged re-entry vehicle launched on 27 December 1983 into an orbit and recovered in the Black Sea after spending about one and a half hours in space. Soviet manned flights with this type of re-usable space vehicle are expected to begin soon.

The Soviet re-usable space vehicle is likely to be smaller than the American Space Shuttle. The Soviet vehicle may weigh about 9 tonnes; the American weighs about 34 tonnes. The US Shuttle can carry a payload of about 15 tonnes into orbit; the payload of the Soviet vehicle will probably be proportionally smaller.

The Soviet and American military could use Space Shuttles for a number of purposes. One obvious use is to transport into space equipment needed for the development of space weapons. They could also be used to transport materials to build 'space battle-stations' equipped with, for example, high-energy lasers. Large structures built in space could be used as platforms for launching weapons at targets on Earth or for stationing troops in space. They could become full-scale military bases.

By 1988, seventy-two US Space Shuttle flights are planned to take place. About one-third of them will carry military payloads. The superpowers are obviously well aware of the new opportunities for military activities in space. The Space Shuttle will certainly greatly facilitate manned activities in outer space. But the tragic explosion of *Challenger*'s launch on 28 January 1986, which killed all seven people on board, will of course delay plans for future Space Shuttle flights, possibly by up to one year.

The USSR has energetically pursued its manned space programme for more than twenty years now and has maintained its lead over the USA. Not only was a Soviet cosmonaut the first man to orbit the Earth but the Soviet Union was the first to launch a man into space.

The Soviets are busily constructing space stations; Soyuz satellites transfer people and goods between Earth and Salyut space stations. The USA intends to build a large space station with materials transported in Space Shuttle flights. Space stations could be used for many military purposes – for the deployment of weapons in space and the stationing of troops. Full-scale military bases in space may be commonplace in the next century.

We usually think only of the USA and the USSR as the military powers in space, but four other countries have launched their own spacecraft, France, China, Japan and India are all space powers. And there are other countries, like Brazil, that have advanced high-altitude rocket programmes in which rockets are fired into the upper atmosphere for scientific experiments. These countries will no doubt eventually develop their own space-launchers and put satellites into space. Many of these satellites will be used to support military activities.

Few people realize the extent to which space has already been militarized. The general public associates satellites mainly with their uses for transmitting live TV pictures around the globe, for relaying intercontinental telephone calls, and other peaceful purposes. But since 1957, when Sputnik began the space age, more than three out of every four satellites launched have been for military use.

So far, more than 2,000 military satellites have been put into space. About 100 new military satellites are launched each year; about 85 by the USSR and about 15 by the USA. The Soviets launch more military satellites than the Americans do because US satellites live, on average, five or six times longer, and carry out more tasks, than their Soviet counterparts. Thus, a typical US Big Bird satellite stays in space for about 200 days, whereas a typical Soviet Cosmos satellite stays up for some 30 days. Big Bird is also more versatile than a Cosmos satellite.

The Americans launch fewer satellites than the Soviets, but the American military are more dependent on them. And the Pentagon is prepared to pay for its space dependency. The US military space budget is running at about $10,000 million a year, five times as much as it was 10 years ago. The military space budget is separate from the budget of the National Aeronautics and Space Administration (NASA), now running at about $7,500 million a year, even though many of NASA's activities have military implications.

Satellites are used by the military for many things – reconnaissance, communications, navigation, early warning of attack, electronic surveillance, ocean surveillance, oceanography, weather forecasting, and so on. The military have become dependent on satellites for intelligence, communications, and early warning of attack. Photographic reconnaissance satellites are the 'eyes' of the military in space; electronic surveillance satellites are its 'ears'; and communications and early-warning-of-attack satellites are its central nervous system.

Electronic surveillance satellites carry equipment to detect and monitor the enemy's radar signals, his radio communications and even his telephone calls. Early-warning satellites are able to detect, among other things, the launching of enemy ballistic missiles very soon after they are fired. The satellites carry sensors that detect the infra-red radiation emitted by the missiles' booster rockets. Immediately they detect enemy missiles or aircraft, the satellites send a signal to a command centre on the ground.

Military space operations are, of necessity, largely automated; in turn, automated warfare will depend on space activities. The most complex military activities in space are those associated with anti-satellite warfare and anti-ballistic missile warfare systems. These weapon systems, if developed and deployed, will have far-reaching consequences for automated warfare.

Anti-satellite warfare

Satellites are so important to the military that it is hardly surprising that both the USA and the USSR are trying to develop weapons to shoot down the other side's satellites in space. Satellites would be so crucial in war that they would be very high priority targets.

A typical Soviet anti-satellite weapon is another satellite, called a hunter-killer or interceptor satellite. The hunter-killer is launched into an orbit that will take it close to the enemy satellite that is its target. When the Soviet hunter-killer is at its closest point to its target it explodes and destroys the enemy satellite. A hunter-killer is normally manoeuvred, by signals from a command centre on earth, towards its target. It then uses its own radar to seek out and make the final approach to the target. The Soviets have been making tests in space with targets and hunter-killer satellites since as long ago as 1967.

Apparently, the last Soviet test of an anti-satellite weapon took place in mid-June 1982. At about 3 p.m. on 18 June 1982, the Soviet satellite Cosmos 1379 was launched from the space-launch centre at Tyuratam. Cosmos 1379 was a hunter-killer; its mission was to seek out and destroy its target, Cosmos 1375, launched on 6 June from Plesetsk. The hunter-killer intercepted its prey during its second orbit of the Earth. When it got close to its target, the killer satellite received a signal from its command centre on Earth. It headed back towards Earth, entered the Earth's atmosphere and burnt up. Its mission had been successfully carried out; if the signal from the command centre had instructed it to, Cosmos 1379 would have exploded and destroyed the target satellite.

Cosmos 1379 was about the fiftieth hunter-killer launched in the Soviet anti-satellite weapon programme. Most of the interceptor satellites were launched from Tyuratam; most of the target satellites were launched from Plesetsk.

In many of these tests the interceptor was exploded with conventional explosives to test its capability to destroy the target satellite. In other tests, the hunter-killer satellite was taken close to the target satellite and, like Cosmos 1379, commanded back to Earth to be burnt up in the atmosphere. There is no point in destroying target satellites unnecessarily.

The Americans have developed a different type of anti-satellite weapon. It consists of a small missile launched into space from a converted F–15 Eagle fighter aircraft in a zoom climb at an altitude of about 20,000 metres. The missile, which may carry a warhead filled with high explosive, is guided close to the enemy satellite. The warhead is then released and guides itself to the target satellite using a sensor in its nose. The sensor is able to pick up and home on the infra-red radiation given off by the enemy satellite. The warhead either explodes when it is close to the target satellite or makes a direct hit on it.

The US Air Force first tested its anti-satellite weapon in January 1984. Although the Americans started late in the anti-satellite arms race they have already caught up with the Soviets, and may already be ahead in the technology.

One direct hit on a satellite from a relatively small warhead travelling fast would disable it. The warhead does not, therefore, need to explode if it is accurately delivered. The Americans are developing a non-explosive manoeuvring warhead for their anti-satellite weapons. The system is called the Miniature Homing Intercept Vehicle. A number of such vehicles could be launched into orbit by a single rocket near a Soviet satellite. Alternatively, the vehicles could be launched by missiles carried by F–15 aircraft. Each of the manoeuvring vehicles carries an infra-red sensor that guides it to the enemy satellite. The vehicle then rams the satellite at high speed and disables it.

The USA is very concerned about Soviet strategic nuclear submarines operating close to its shores. Ballistic missiles fired from these submarines could reach their targets in minutes rather than the half-hour or so taken by intercontinental ballistic missiles fired at the USA from the Soviet Union. Because of the threat from Soviet submarines close to US shores, the US Air Force intends to arm several units with anti-satellite weapons capable of attacking the low-orbit Soviet satellites supporting Soviet strategic nuclear submarines.

Yet another type of anti-satellite weapon being actively considered consists of a rocket carrying a payload of a large number of ball-bearings or small pellets. The idea is to discharge them as a cloud in the path of an enemy satellite. The target satellite would be so damaged by flying through the cloud that it would be disabled.

The anti-satellite weapons that have been developed so far by the USA and the USSR are effective against enemy satellites in low orbits, 300 to 1,500 kilometres above the Earth. But neither side has yet managed to knock out satellites in very high orbits. This they would very much like to do. In particular, they would like to be able to attack the other side's communications and early-warning-of-attack satellites in geostationary orbits, about 35,000 kilometres above the Earth. (These satellites move so as to hover above the same area of the Earth's surface.) To attack satellites in geostationary orbits would require high-energy laser weapons of the type envisaged for a 'Star Wars' defence.

Star Wars

President Reagan took almost all Americans by surprise when he suggested that America try to defend itself from a full-scale nuclear attack by Soviet ballistic missiles. The President unveiled his 'Star Wars' plan in a speech on 23 March 1983. 'I call upon the scientific community who gave us nuclear weapons,' the President said, 'to turn their great talents to the cause of mankind and world peace: to give us the means of rendering these nuclear weapons impotent and obsolete.' President Reagan's plan is to use very exotic weapons, like high-energy lasers, on 'space battle-stations' as part of a defence system with the job of destroying in flight Soviet ballistic missiles attacking the United States. That the President can call for an armada of 'space battle-stations' to defend the USA against nuclear attack brings home the extent to which space has become militarized.

'Star Wars' is mainly an anti-ballistic missile programme designed to intercept and destroy ballistic missiles in flight. But the technology required for this task is essentially the same as that required for effective anti-satellite warfare systems. Anti-ballistic missiles and anti-satellite warfare technologies go hand-in-hand. Anti-ballistic missile technology is of interest to us here also because, if it could be made effective, particularly against tactical ballistic missiles, it would have far-reaching consequences for the automated battlefield.

Anti-ballistic missile systems are not new. In the 1960s, both the USA and the USSR developed such systems, using interceptor missiles of very high acceleration armed with nuclear warheads. Complex radars are used to detect and track the incoming enemy warheads. An anti-ballistic missile is fired and guided by the radars towards an incoming enemy warhead. When the anti-ballistic missile gets close to its target, its nuclear warhead explodes to destroy the enemy warhead.

In 1968, the Soviet Union started operating such an anti-ballistic missile system around Moscow using Galosh interceptor missiles. In 1974, the Americans built a similar system around an intercontinental ballistic missile site in North Dakota. The Soviets are still operating, and modernizing, their system around Moscow. But the Americans soon lost faith in theirs and dismantled it in 1975.

These early efforts showed how very difficult it is to defend even a small area against a ballistic missile attack. But new technology, particularly the ability to collect and rapidly analyse huge amounts of data, has given rise to claims that an effective anti-ballistic missile system may become feasible in, say, ten or twenty years from now.

The protection of an area attacked by a large number of ballistic missiles would be conceivable only if there were several layers of defence, each able to destroy a high proportion of the attacking missiles. The first layer of defence would have to destroy most of the attacking warheads soon after they were fired, either from land-based silos or from submerged submarines. In other words, if a ballistic missile defence system were to be successful, most of the enemy missiles would have to be destroyed while their booster rockets were still firing. Enemy missiles that survived the first layer of defence would be attacked in space by weapons in the second layer of defence. The warheads of any missiles that survived the second layer would be attacked, by weapons in the third layer, as they left space and re-entered the Earth's atmosphere.

In the proposed American system, the weapons in the first layer of defence would include high-energy lasers deployed on space battle-stations. The second layer would consist mainly of other space battle-stations carrying high-energy lasers to attack Soviet warheads in space. The third layer of ballistic missile defence would consist of anti-ballistic missiles fired from the ground.

Laser beams are beams of light. When you switch on the electric light, light streams out of the bulb in all directions, lighting up the whole room. But light from the device called a 'laser' goes in one direction only, travelling in a narrow beam. An enormous amount of energy can be concentrated in a laser beam, enough to cut a hole through, for example, a steel block several centimetres thick. And this is why high-energy laser beams are of great interest to the military.

The USA has already tested some laser weapons on Earth. For example, in 1978 the US Navy destroyed a missile in flight with a laser weapon. In 1983, the Airborne Laser Laboratory, operated by the US Air Force, showed that Sidewinder air-to-air missiles can be shot down in flight by laser weapons, carried by an aeroplane, at ranges of between eight and sixteen kilometres. The airborne laboratory shot down all five Sidewinder missiles fired at it from another aircraft. It is believed that new Soviet battleships are being equipped with high-energy laser weapons to defend them against attacks by, for example, enemy cruise missiles.

But none of the lasers so far tested are anywhere near powerful enough to destroy ballistic missiles in space. Great efforts are being made, though, in both the USA and the USSR, to increase the power of laser weapons. American military scientists are developing a large mirror, four metres in diameter, to steer a high-energy laser beam in space. And in the Talon Gold project, technologies for aiming laser weapons at enemy ballistic missiles are being investigated.

If a high-energy laser beam hit a ballistic missile or its warhead it would probably damage it enough to make it ineffective. It would also destroy an enemy satellite in orbit, even if the satellite were deep in space. Lasers are, therefore, crucial components of many space weapons.

Soviet land-based ballistic missiles would be fired a long way from American, or NATO, territory. These missiles could be attacked at an early stage in their flight only from space. This is why the current proposal for an anti-ballistic missile system is primarily a space-based programme, hence the name 'Star Wars'.

The fact that the Star Wars concept depends on destroying most Soviet missiles just after they are fired distinguishes it from all previous anti-ballistic missile concepts. The earlier systems were designed to attack all the enemy warheads at the other end of their flights, as they re-entered the Earth's atmosphere from space. The earlier anti-ballistic missiles, moreover, were armed with nuclear weapons. Star Wars will, it is planned, use non-nuclear methods to destroy enemy missiles and warheads.

An American anti-ballistic missile system would have very little time to attack Soviet ballistic missiles as they rose through the atmosphere. The booster stage of a SS–18 Soviet intercontinental ballistic missile, for example, burns out after 300 seconds when the missile is outside the Earth's atmosphere, at a height of 400 kilometres. New missiles will have even shorter boost phases. The booster of the American MX missile, for example, stops burning after 180 seconds, when the missile is outside the atmosphere at an altitude of about 200 kilometres. And it is quite feasible to develop a booster that would burn for only 40 seconds, when the missile was at an altitude of only 80 kilometres.

Strategic ballistic missiles fired from submarines or tactical ballistic missiles would be more difficult to attack than land-based strategic ballistic missiles. Submarine-launched and tactical ballistic missiles have relatively short flight times from their launch points to their targets. Typically they would have flight times of less than ten minutes, compared with about thirty minutes for a land-based intercontinental ballistic missile travelling 10,000 kilometres or so. Mobile missiles, moreover, would be launched from unpredictable points. The silos of fixed land-based ballistic missiles, on the other hand, can be pin-pointed.

A typical modern strategic ballistic missile carries, not a single warhead, but many warheads. The warheads are carried on a manoeuvrable vehicle called a 'bus'. Once in space, the bus releases the warheads, one after the other, each along a slightly different trajectory. Each warhead can, therefore, attack a separate target; the targets can be hundreds of kilometres apart.

A single missile can carry up to fourteen warheads. But it can also carry a large number of decoys, such as light-weight balloons, that in space would travel along ballistic trajectories indistinguishable from those of real warheads. This is another bugbear for anti-ballistic missile systems. If enemy warheads survive the boost phase and go into space, each might proliferate into a hundred or more similar objects. All would have to be destroyed to be sure of getting the warheads.

It is because of the very short time available to attack the missiles before they get into space, and the importance of destroying them before they release their warheads, that the weapons being mainly considered for Star Wars are high-energy lasers on space battle-stations. Laser beams travelling at the speed of light, 300,000 kilometres a second, are the most feasible weapons for the job.

Laser beams on space battle-stations could also be used to attack warheads in space. But another system has been suggested by some American military scientists for the second layer of a ballistic missile defence system. This also relies on high-energy laser beams deployed, not in space, but on top of a high mountain, say 5,000 or more metres above ground level. A large mirror would be launched into an appropriate space orbit to reflect the laser beam coming from the mountain top and to aim the beam at the enemy missile. A laser weapon capable of producing enough energy to damage an enemy missile sufficiently to make it useless would be of great weight. To launch such a weight into space would be an expensive business. Putting the laser on a mountain top would be easier. Earth-based lasers would also be easier to repair and maintain than space-based ones.

A major problem with using high-energy lasers as weapons is that when the light in a laser beam travels through air it does not stay in a narrow beam; the beam tends to broaden out and is, therefore, capable of focusing less energy on a small area. Broad laser beams are much less efficient at damaging missiles than narrow ones.

One of the attractions of putting high-energy laser weapons into space is that there is no air in space to broaden the beam, and so the beam remains relatively well focused. Putting the laser weapon on a high mountain, above much of the atmosphere, is another way of reducing the problem of transmitting a narrow beam through air.

Star Wars advocates believe that about fifty space battle-stations armed with high-energy lasers would be enough to protect the entire USA from a mass attack by Soviet strategic ballistic missiles. Other scientists say that many more than fifty space battle-stations would be required.

Each space battle-station would carry sensors and computers to detect, identify, and track enemy ballistic missiles as they were launched. Equipment on the space station would aim a laser beam at each enemy missile and fire it at the appropriate moment.

Another possible anti-ballistic missile weapon for a space battle-station is the electromagnetic rail gun. This futuristic weapon would use extremely strong magnetic fields to accelerate shells along metal guides and out through space. The shells would be aimed at enemy warheads. Target acquisition and aiming would probably be done using a laser system and the shell would be equipped with terminal guidance. If one of the shells from an electromagnetic rail gun hit an enemy warhead it would destroy it by the energy of the impact, like a bullet would. But to get to an enemy warhead in the short time available the shell would have to be accelerated to a very high velocity indeed.

An experimental rail gun has been built at the University of Texas. It has succeeded in accelerating small shells, weighing 85 grams, to high velocities but is a far cry from an effective electromagnetic rail gun for use against ballistic missiles. This would need to fire shells weighing several thousand grams and fire many of them in a short time. The US Army believes it can successfully develop a rail gun and is spending much money on research and development.

The space battle would be controlled from the ground by the Consolidated Space Operations Center at Colorado Springs. The Center's main computer would receive information from early-warning-of-attack satellites and, if necessary, activate the space battle-stations. The space battle would then begin.

Much research is now being done in the USA on technologies for space-based anti-ballistic missile systems. President Reagan wants to give $26,000 million for this research over the next five years.

Star Wars enthusiasts believe that an effective system could be deployed in the 1990s or early next century. Other scientists doubt that the technologies to defend the USA effectively against a full-scale attack by Soviet strategic ballistic missiles can be developed. There is more agreement about the feasibility of defending smaller areas against an attack by ballistic missiles, strategic or tactical.

The technologies being developed are designed to attack enemy ballistic missiles and their warheads. They may also be effective against enemy aircraft flying at high altitudes. But they would be ineffective against cruise missiles flying at low altitudes.

Large space battle-stations permanently deployed in space would, of course, be extremely vulnerable to attack. They would have to be defended and, given their huge cost, very well defended. The need to defend space battle-stations will encourage the development and deployment of space-based radars, probably powered by solar cells. These radars would give warning of an enemy attack on a space station in time to allow the Space Operations Center to manoeuvre the station to try to evade the attack.

The domination of space would be a necessary precondition for reliable large-scale ballistic missile defence. But the domination of space requires effective anti-satellite warfare systems able to destroy enemy satellites in orbit. And this is why anti-ballistic missile and anti-satellite warfare systems are bedfellows.

Chapter 8
Advanced Weapons in the Third World

The time is mid-1995. Imagine the following events taking place in Saudi Arabia. An oil well is blown up by Saudi rebels. This is the most serious of a number of recent violent incidents. Saudi King Fahd Ibn Abdul Aziz, convinced that a serious rebellion is under way, asks the American President to send to Saudi Arabia units of the US Rapid Deployment Force. The President agrees. But the presence of the US marines involves the USA in the Saudi civil war.

An American military transport plane is shot down as it approaches Riyadh airport. All 300 troops of the Rapid Deployment Force on board are dead. Pentagon sources say that the plane was hit by a surface-to-air missile fired by rebels at a range of several kilometres.

There is no doubt that the rebels are using precision-guided munitions – missiles that can hit targets with great accuracy. Such rockets were used to attack the oilfield as well as bring down the transport aircraft. The weapons are the latest automated weapons. Once fired, they seek out, identify and attack their targets automatically, without any further instructions from the launcher, so-called 'fire-and-forget' missiles. The rebels are obviously launching the missiles at relatively long ranges. This makes it very difficult for Government troops to catch them.

The weapons used by the rebels seem to be of Soviet or Czechoslovakian origin. They come through Iraq and are smuggled into Saudi Arabia either across the border or by sea. The use of Soviet weapons involves the USSR in the Saudi conflict. Now both superpowers are known to be involved in a war in this unstable Third World region. This involvement could escalate to a conflict between the superpowers.

The superpowers become effectively the guarantors of the survival of their clients (just as the USA is the guarantor of Israel's survival and the USSR is the guarantor of Syria's survival). The USA cannot risk the loss of credibility that would follow if the Saudi rebels defeated the Saudi Government. In its turn, the USSR cannot risk the loss of face that it would suffer if the rebels were defeated. Both superpowers have to maintain their credibility. Not to do so would be to risk losing allies around the world. This superpower rivalry clearly carries the seeds of escalating conflict between the superpowers.

The weapons used by the rebels are obviously very powerful. Given money and contacts, the Saudi experience shows that it is relatively easy for sub-national groups to get hold of autonomous weapons of high fire-power. Access to the most sophisticated modern weapons increases considerably the probability of low-level, guerrilla war. They also increase the risk of strategic nuclear war between the USA and the USSR. Terrorism and the risk of strategic nuclear war are connected.

Imagine what would happen if the Saudi rebels got hold of some plutonium and made a nuclear explosive. Governments known to be sympathetic to rebel causes, like the Libyan and Cuban Governments, now (in 1995) have stocks of plutonium that may be given away or stolen.

Rebel groups, like the Saudi rebels, know that they would put the cat among the pigeons by exploding a nuclear weapon, or even threatening to explode it, in Washington or Moscow. They could easily smuggle the components in in suitcases. A nuclear explosion in the USA or the USSR at a time of serious international crisis could easily escalate to a nuclear world war. This is why there is a direct relationship between modern weapons in the hands of non-governmental groups and a strategic nuclear war between the superpowers.

But do rebels have the expertise to make a nuclear weapon? Any significant rebel or terrorist group has the expertise. If it does not have it within its ranks, it could buy it. There are many thousands of people either with experience of designing nuclear weapons or with the knowledge of how they work; people who, for example, have worked with nuclear weapons in the armed forces of the nuclear-weapon powers.

Terrorism

Terrorism is as old as history. The era of modern terrorism began a hundred years ago in Tsarist Russia with a surge of the type of terrorism described by Karl Marx in 1848. 'Only one means exists to shorten the bloody death pangs of the old society and the birth pangs of the new society, to simplify and concentrate them – revolutionary terrorism.' But since the Second World War new types of terrorism have emerged.

Many contemporary terrorists begin with motives we can understand and often support – a desire for social justice, concern for the environment, national liberation and so on. Others are religious fanatics. But the knowledge that they have wide support for their values often gives would-be terrorists the belief that they can defy the rule of law to gain their ends and they begin to do so.

In the words of Ezekiel Drorr, an Israeli expert on terrorism, today's typical terrorist is an 'idealist who despairs of achieving his values by normal means and, therefore, becomes a fanatic who we define as an idealist whom we don't like, someone who uses tools and instruments that we regard as immoral. Modern society provides very few legitimate roles for persons who centuries ago would have become saints, explorers, or adventurers.'

Terrorist activity has rapidly increased in recent years. According to official US statistics, the number of international terrorist incidents increased from about 120 in 1968 to 652 in 1984. And in just the first quarter of 1985 there were 200 terrorist incidents. These incidents include kidnapping, hostage taking and barricading, bombing, armed attack, hijacking, arson and shootings.

In the five years 1979 to 1983 inclusive, about 2,100 people were killed and 4,400 wounded in international terrorist incidents. About 44 per cent of terrorist attacks in this period were against diplomatic buildings and diplomats, and 12 per cent involved military establishments and personnel. The most common terrorist activity is bombing. In 1983, typical of recent years, of 500 international terrorist incidents, 262 were bombings. The next most common incidents were armed attacks; in 1983, there were 68, including 36 assassination incidents. Of the incidents in 1983, 37 per cent took place in Western Europe, 26 per cent in Latin America, 23 per cent in the Middle East, 8 per cent in Asia, 3 per cent in Africa, 2 per cent in North America and 1 per cent in Eastern Europe and the USSR.

So far, terrorists have been relatively unimaginative in their actions and their activities have been relatively well contained. They have confined themselves to the use of conventional weapons. But we must expect this pattern to change in the future. Existing terrorist groups are small and, therefore, the scientific and technical skills in any one group is limited. But, in the future, some groups may combine their resources and thus contain enough skill to be able themselves to produce unconventional weapons, like nerve gases or nuclear explosives. Or extremist terrorist groups may team up with technologically advanced extremist states and be given weapons of mass destruction by these countries. Very probably, leaders of terrorist groups have already at least thought about using nuclear, chemical or biological weapons. Presumably, they have until now decided that killing, or threatening to kill, large numbers of people indiscriminately would not further their ends. But, if access to weapons of mass destruction becomes easier, this attitude may well change.

In the coming years, the nature of terrorism will almost certainly change and become much more aggressive. The extent and frequency of terrorist activity is likely to escalate. This will happen mainly because of the increase in the world's population and, consequently, in urbanization. The predictions of population increase are staggering. Demographers tell us that, unless catastrophes occur to prevent it, the world population will inevitably reach roughly 11,000 million by the year 2030 or so. (Today's world population is about 4,600 million.) This population is predicted even though rather optimistic assumptions are made about increases in the use of birth control methods. Of these people, roughly 9,000 million will live in the Third World. And of this Third World population, about 6,000 million will live in cities.

We can, therefore, look forward to vast metropolises, cities containing 30, 40, 100, or more millions of people. In most of these cities, a large fraction of the population will live in sub-standard (shanty) housing, be illiterate, have no access to health services and so on. The potential for urban guerrilla violence in these unimaginably bad conditions is obvious. The pattern of future terrorism will largely depend on the weapons available to terrorists. And this, in turn, will depend on how modern weapons spread worldwide, particularly to Third World countries. We will now discuss how the world's arsenals are likely to grow.

Global militarization

A main characteristic of the growth in the world's arsenals over the past two decades or so is the widening gulf between the military might of the great powers and that of the rest of the world. Although most countries are increasing their military potential steadily, the superpowers are increasing theirs at a faster rate than other countries. And the gap is likely to widen.

The reason why this hierarchy of military power has developed is not difficult to find. The superpowers have invested large sums of money in their arsenals since the Second World War. In the past twenty years, for example, the USA and the USSR have spent, in 1980 dollars to correct for inflation, about $6,000,000 million on their military establishments. In 1984, the USA and the USSR together spent about $600,000 million on their militaries. These two countries alone spend about 70 per cent of total world military expenditure. The superpowers spend about $60,000 million a year on military research and development, about 85 per cent of global spending on military research and development. In these two countries, about 400,000 scientists and engineers are working full-time on improving the performance of existing weapons and developing new weapons.

No country other than the superpowers is able to invest such huge resources in military science or in procuring weapons. Consequently, none can compete with the superpowers in the sophistication of the weapons in their arsenals or in the size of their arsenals.

Nevertheless, more and more countries are developing a modern military technological base. At the end of the Second World War, only a small handful of countries, including the USA, the USSR, the UK, Canada, and Sweden, developed and produced major weapons. Today, a relatively large number of countries, both industrialized and developing, do so. Industrialized countries with significant defence industries include: Austria, Belgium, Canada, France, Sweden, the UK, and Yugoslavia. These countries produce at least modern combat aircraft and missiles. Czechoslovakia, Finland, the Netherlands, Poland, Romania, Spain and Switzerland produce modern combat aircraft.

Some developing countries also produce modern weapons. Argentina, Brazil, China, Egypt, India, Indonesia, Israel, North Korea, South Korea, Pakistan, the Philippines, South Africa and Taiwan produce modern combat aircraft. Except for Indonesia, North Korea, South Korea and the Philippines, they also produce missiles. Some of these developing countries, like Argentina, Brazil, China, Israel and South Africa, have established relatively large defence industries and do significant military research and development. They also sell some of the weapons they make to other countries.

The proliferation of weapon-producers inevitably means that sophisticated weapons are spreading worldwide. All producers are anxious to keep up with the latest technologies. Details of the design of modern weapons spread because of the global arms trade. Worldwide, about $200,000 million worth of weapons are currently produced a year. About $50,000 million worth of these weapons are exported. In a typical year, the USA and the USSR account for about 70 per cent of the global arms trade, France accounts for about 10 per cent, and the UK, Italy, and West Germany account for about 4 per cent each. Weapons exported by these countries go to all the world's continents, over two-thirds going to Third World countries. The Middle

East is the biggest weapons importing region in the Third World, accounting for about 45 per cent of imports in a typical recent year. Asia and Africa each import about 22 per cent of the Third World total. And Latin America accounts for about 10 per cent. Well over 90 per cent of the arms trade is approved, and licensed, by governments. Private arms merchants have a relatively small share of the market, although the trade is so huge that the absolute amount of money involved in private arms deals is large.

There are many examples of how weapons sales have boomeranged and the weapons been turned against the sellers. One example is the Falklands War between the UK and Argentina. Before the war Britain sold many major weapons to Argentina. And many of the 250 British servicemen killed in the war were killed by some of those weapons. *HMS Ardent*, for example, was sunk by Aermacchi MB339A jet aircraft; 24 of the crew of 170 were killed and 30 injured. The Aermacchis are powered by British Rolls-Royce engines. And many parts of the Exocet missiles used by the Argentinians are British-made. Increasingly, weapons sold abroad will boomerang in the hands of sub-national groups.

Because the global arms trade is such a cut-throat business, governments that export weapons are so anxious to sell their wares that they will often sell weapons abroad before they acquire them for their own arsenals. Consequently, the most sophisticated weapons and their supporting technologies are available in the global arms markets. In today's world, if you want to buy the latest weapons, including missiles with considerable fire-power, someone will be prepared to sell them to you. And the world's arsenals now contain so many advanced conventional weapons that any determined group, of, for example, terrorists, can acquire the weapons it wants.

The spread of nuclear weapons

If sophisticated, very destructive conventional weapons are widely available, what about nuclear weapons? The ability to produce nuclear explosions is spreading and, unless nuclear technology is controlled, we must expect nuclear capabilities to spread further.

There are now five established nuclear-weapon powers: the USA, the USSR, China, France and the UK. We are sure that these powers have nuclear weapons; we know quite a lot about the size of their nuclear arsenals and the characteristics of the nuclear weapons in them. Each of these nuclear-weapon powers explodes a nuclear weapon from time to time; on average, there is one nuclear test a week. Since the first nuclear-weapon was exploded in 1945, at least 1,450 nuclear explosions have been carried out.

India exploded a nuclear device in 1974. Just before she was killed, Indian Prime Minister Indira Gandhi said that India had no nuclear weapons and so presumably the weapon exploded was the only one India had produced. India could, of course, produce other nuclear weapons very quickly if the political leadership decided to do so.

Most people believe that Israel has nuclear weapons. From time to time, the American Central Intelligence Agency announces that Israel has about twenty-four nuclear weapons. And many people believe that South Africa has them too. We are not sure about Israel or South Africa; neither country is definitely known to have exploded a nuclear device (although there is some evidence from a US Vela satellite that South Africa has exploded one). Nor have the political or military leaders of South Africa or Israel admitted to having nuclear weapons. It is part of the policies of these two countries to keep the rest of the world guessing about their nuclear capabilities.

A crucial question is: could a sub-national group of, for example, terrorists or even criminals produce a nuclear weapon?

The basic nuclear weapon is the fission weapon. A fission chain reaction is used to produce a very large amount of energy in a very short time and, therefore, a very powerful explosion. The fission occurs in a heavy material, specifically uranium or plutonium. The nuclear weapons built so far have used the isotopes uranium–235 and plutonium–239. Fission occurs when a neutron enters the nucleus of an atom of one of these materials, which then breaks up, or fissions. When fission occurs a large amount of energy is released, the original nucleus is split into two radioactive nuclei (the so-called fission products), and two or three neutrons are released. The neutrons can be used to produce a self-sustaining chain reaction. A chain reaction will occur if at least one of the neutrons released in each fission produces the fission of another atom of uranium or plutonium, as the case may be.

There exists a critical mass for uranium–235 and plutonium–239 that is the smallest amount of the material in which a self-sustaining chain reaction, and hence a nuclear explosion, will take place. A critical mass of, for example, a bare sphere of plutonium–239 would weigh about 10 kilograms and be about the size of a small grapefruit. Using a technique called implosion, in which a plutonium sphere is surrounded with conventional high-explosive lenses to compress a mass slightly less than critical to a mass which is slightly greater than critical, a nuclear explosion can be achieved with about two or three kilograms of plutonium. A sphere of plutonium of this weight would be smaller in volume than a tennis ball.

In the nuclear weapon that destroyed Nagasaki, on 9 August 1945, about eight kilograms of plutonium metal were used. The plutonium was surrounded by a tamper that had two purposes. First, it was to reflect back into the plutonium some of the neutrons that escaped from the surface of the core, allowing some reduction in the mass of plutonium needed to produce a nuclear explosion. The second purpose of the tamper was more important. Because the tamper was made of uranium, a very heavy material, its inertia helped to hold the plutonium together during the explosion to prevent the premature disintegration of the plutonium and thus obtain greater efficiency. The plutonium core was surrounded by chemical high explosives arranged as explosive lenses focused on the centre of the plutonium sphere. When the lenses were detonated the sphere was compressed uniformly by the implosion. The compression increased the density of the plutonium so that the slightly-less-than-critical mass was made slightly more than critical. The bomb then exploded.

The complete detonation of eight kilograms of plutonium

would produce an explosion equivalent to the explosion of 144,000 tonnes of a high explosive like TNT. (Because 1,000 tonnes is called a kiloton, this is equal to 144 kilotons.) The eight kilograms of plutonium in the Nagasaki bomb produced an explosion equivalent to that of twenty-two kilotons of TNT. The efficiency of the Nagasaki bomb was, therefore, 15 per cent. Modern nuclear weapons are much more efficient.

In the nuclear weapon that destroyed Hiroshima on 6 August 1945, a different design was used. A sub-critical mass of uranium–235 was fired down a gun barrel, actually an old naval gun barrel, into another sub-critical mass of uranium–235 placed in front of the muzzle of the gun. About sixty kilograms of uranium–235 were used in the Hiroshima bomb to produce an explosion equivalent to that of about thirteen kilotons of TNT. The 'gun' design is cruder than the implosion method and is rarely used in modern nuclear weapons. But a sub-national group may find it easier to construct than the implosion method.

Nuclear design has, in fact, advanced considerably since 1945. The Hiroshima bomb, for example, was about 3 metres long, 1.5 metres in diameter and 4,500 kilograms in weight. A modern American nuclear warhead weighs about 100 kilograms and explodes with an explosive power equal to that of 350,000 tonnes of TNT. Some modern nuclear weapons, such as a nuclear 155 mm artillery shell, have relatively small physical dimensions.

Crude nuclear explosives

A sub-national group could make an effective nuclear explosion using a design much cruder than the Nagasaki one. In the words of Mason Willrich and Ted Taylor (the latter a veteran nuclear-weapon designer):

> Under conceivable circumstances, a few people, possibly even one person working alone, who possessed about 10 kilograms of plutonium oxide and a substantial amount of

> chemical high explosive could, within several weeks, design and build a crude fission 'bomb'. By a 'crude fission bomb' we mean one that would have an excellent chance of exploding with the power of at least 100 tonnes of chemical high explosive. This could be done using materials and equipment that could be purchased at a hardware store and from commercial suppliers of scientific equipment for student laboratories. A crude fission bomb, as we have described it, might yield as much as 20,000 tonnes of explosive power – the equal of the Nagasaki A-bomb.

According to the Office of Technology Assessment of the US Congress:

> a small group of people, none of whom have ever had access to the classified literature, could possibly design and build a crude nuclear explosive device. They would not necessarily require a great deal of technological equipment or have to undertake any experiments. Only modest machine-shop facilities that could be contracted for without arousing suspicion would be required. The financial resources for acquisition of necessary equipment on open markets need not exceed a fraction of a million dollars. The group would have to include, at a minimum, a person capable of searching and understanding the technical literature in several fields and a jack-of-all-trades technician. Again, it is assumed that sufficient quantities of fissile material have been provided.
>
> The actual construction of even a crude nuclear explosive would be at least as difficult as the design itself. The small non-national group described above would probably not be able to develop an accurate prediction of the yield of the device. The device could be a total failure, because of either faulty design or faulty construction. Here again, a great deal depends on the competence of the group; if it is deficient, not only is the chance of producing a total failure increased, but the chance that a member of the group might suffer serious or fatal injury would be quite real. However, there is a clear possibility that a clever and competent group could design and construct a device which would produce a significant

nuclear yield (i.e. a yield *much* greater than the yield of an equal mass of high explosive).

Crude designs, much simpler than the 1945 Nagasaki design, are likely to produce nuclear explosions equivalent to the explosion of between 100 and 1,000 tonnes of TNT. They might yield several thousand tonnes, but are unlikely to yield more than 10,000 tonnes.

The crudest device could be very crude indeed and could be simply constructed by competent technicians (or a competent technician working alone) from, say, a sub-critical mass of plutonium oxide. It must be remembered that an explosion equivalent to that of a few tens of tonnes of TNT would completely destroy the centre of a large city. The largest conventional bomb used in the Second World War, called the 'earthquake' bomb, used about ten tonnes of TNT.

A 100-tonne explosion on the surface would produce a crater about 30 metres across. A nuclear explosion of this power would produce about 500 rems (radiation units) of prompt gamma-radiation at ranges in open air of up to 300 metres; 500 rems would kill about half of the people exposed to it. It would also produce about the same dose of prompt neutrons at about 450 metres and about 500 rems of fall-out radiation in the first hour to those within a few hundred metres downwind. An explosion equivalent to 1,000 tonnes of TNT would devastate about five square kilometres.

The high explosives and detonators needed for a nuclear explosive based on the Nagasaki design of the Second World War are now commercially available. The information needed to design such a nuclear device is also widely available. A small country might well use this type of design to produce nuclear weapons in whose effective operation it could be confident without the need for nuclear testing. Amory Lovins, an American physicist who has made a special study of the proliferation of nuclear weapons, describes it as 'not a clever technology (simpler means could produce better results) but it is basic and reliable, and has been the subject of several amateur design exercises'.

A large terrorist group would probably have within its ranks enough competent people to produce a low-technology, 1945-vintage, nuclear weapon. If it did not have this competence it could almost certainly buy it. Any competent nuclear physicist should be able to design such a device without access to classified literature. And there are a large number of people around who have had experience in nuclear-weapon design and construction at some time in their careers.

Availability of plutonium

Given that the design of nuclear fission weapons is no longer secret, how easy is it for governments and sub-national groups to get hold of the nuclear material that they would need to make nuclear explosives? As we have seen, the material needed for nuclear explosives is either plutonium or enriched uranium.

Plutonium does not exist in nature but it is produced as an inevitable by-product in nuclear reactors. The fuel in a reactor is normally uranium, an element occurring naturally in many places. As the fuel is burned up in the reactor, nuclear fission occurs in the uranium. The heat produced by the fission is used to change water into steam. The steam is used to drive a turbine to produce electricity. But as the uranium is used up, plutonium is produced.

Any nation which has nuclear-power reactors is accumulating plutonium on its territory which could be used in nuclear weapons. A typical nuclear-power reactor generates about 1,000 million watts of electricity (MWe), and produces each year about 250 kilograms of plutonium.

In 1984, there were about 350 nuclear-power reactors generating about 210,000 MWe in about twenty-five countries. About fifteen of these power reactors, generating about 6,000 MWe, are in the Third World. By 2000, according to conservative estimates, the world's nuclear capacity will have increased to some 600,000 MWe. Of this capacity about 50,000 MWe may be in developing countries. Developing countries will then be producing about 7 per cent of their electricity by nuclear power, compared to about 30 per cent in the industrialized countries.

The Third World countries now operating nuclear-power reactors include Argentina, Brazil, India, Pakistan, South Korea, and Taiwan. Nuclear-power plants are under construction in these six countries and in Cuba, Mexico, and the Philippines. In at least four other Third World countries – China, Egypt, Libya, and Turkey – nuclear-power plants are in the final planning stage. If capital is made available, perhaps another ten developing countries may start constructing nuclear-power plants by the end of the century.

A world nuclear capacity of 600,000 MWe in the year 2000 will produce about 150,000 kilograms of plutonium annually, theoretically enough to produce about 15,000 nuclear explosives, or about one every twelve working minutes. Of this plutonium, about 12,000 kilograms a year will be produced in, and owned by, Third World countries.

Clandestine production of nuclear weapons

Access to a large nuclear-power reactor is not necessary to produce plutonium for nuclear weapons. A small reactor would do the job. A reactor with a power of, say, about 40 MWe could produce enough plutonium a year for two nuclear weapons of the Nagasaki size. The components for such a reactor can be easily, and secretly, obtained on the open market.

The reactor and a small chemical plant to remove the plutonium from the reactor fuel elements could be hidden in a building or underground. It was in this way that Israel acquired, from France, and operated the reactor at Dimona to produce the plutonium for its nuclear-weapon force.

It is because of this possibility of clandestine production that we do not know for certain which governments, or sub-national groups, have nuclear weapons and which do not.

Reprocessing

Plutonium is removed from spent reactor fuel elements by a chemical process, called reprocessing, in which the fuel elements to be reprocessed are taken to a reprocessing plant where they are dissolved in acid. The solution so formed is then chemically separated into three parts. One contains the plutonium, another contains unused uranium and the third contains the fission products. Because of the danger that plutonium will get into the wrong hands and be used to produce nuclear explosives, many believe that it should be left in spent reactor fuel elements, which should be stored in permanent disposal sites. Spent reactor fuel elements are so radioactive as to be self-protecting. It would be extremely hazardous for people to handle them without large remote-handling equipment. But after the fuel elements have been reprocessed and the plutonium has been separated from the radioactive fission products, the plutonium is relatively accessible.

Reprocessing is undertaken to acquire plutonium either for use in nuclear weapons or as fuel for breeder reactors. All the nuclear-weapon powers (the USA, the USSR, the UK, China and France) have large military reprocessing plants. In addition, some countries (including some that do not have nuclear weapons) are considering building large commercial reprocessing plants.

Most reprocessing plants were planned when the economics of plutonium fuel (for breeders) were thought to be favourable. This is, however, no longer the case. Nevertheless, several countries plan to build reprocessing plants.

Looking ahead to the year 1990, the amount of nuclear power in the countries outside the communist area will probably be about 303,000 MWe; this amount of nuclear capacity will generate about 7,400 tonnes of spent fuel per year. In 1990, four countries plan to have large commercial reprocessing plants. France plans to be able to reprocess (commercially) 2,400 tonnes per year of spent fuel, compared with its current capacity of 800 tonnes; Japan intends to have a reprocessing capacity of 1,700 tonnes per year, compared with the current

200 tonnes; the USA may be able to reprocess (commercially) 1,500 tonnes of spent fuel per year compared with zero capacity today; and the UK plans a commercial reprocessing capacity of 1,300 tonnes per year compared with zero capacity today.

In addition, in 1990, Italy, West Germany, Belgium, Brazil, and India plan to have some reprocessing capacity – 510, 390, 300, 300, and 1,000 tonnes per year, respectively. India now has a reprocessing capacity of about 100 tonnes a year; the other countries have little or no current capacity. Countries like Argentina, which has a small experimental reprocessing plant, may also decide to reprocess on a significant scale.

Incidentally, although less than 5 per cent of the planned reprocessing capacity will be Third World countries – Argentina, Brazil, and India – these countries will, nevertheless, have the capacity to produce about 4,000 kilograms of plutonium a year, enough for about one nuclear explosive a day.

If breeder reactors are uneconomic, what is the purpose of the reprocessing plants? There are several possible answers to this question. Some countries may want access to reprocessing facilities to acquire the option to produce nuclear weapons if they should ever take the political decision to do so. Perhaps this is a main motivation for Argentina, Brazil, and India. The French breeder-reactor programme is almost certainly related to the French nuclear-weapon programme. French plans involve the deployment of many new nuclear warheads over the next few years – so many that the military nuclear facilities are unable to produce enough plutonium. France hopes, therefore, to use the plutonium produced by its breeder reactors for military purposes.

Countries anxious to get into, or improve their performance in, the nuclear export markets may seek reprocessing capabilities to be able to provide more nuclear services than their competitors. Potential exporters may feel they have a greater chance of, for example, selling reactors abroad if they can include the reprocessing of customers' spent reactor fuel as part of the contract. Some countries, like West Germany, have gone even further and agreed to construct reprocessing facilities on the customer's own territory.

Nuclear transportation

Decisions to reprocess on a large scale would inevitably involve the international transport of nuclear material in a massive way. This would occur because commercial reprocessing is a large, capital-intensive, industrial undertaking and only six or eight industrialized countries are likely to engage in it. In, say, the year 2000, spent reactor fuel elements will be transported from the thirty or forty countries operating nuclear-power reactors to the six or eight countries operating the reprocessing plants. Separated plutonium will normally be transported back to the countries owning it. About 16,000 tonnes of spent fuel elements will be available for reprocessing a year; about 160 tonnes of plutonium will be separated. These operations will involve transporting large quantities of nuclear material by all forms of transport – by road, rail, sea, inland waterways, and air.

Nuclear transports are likely targets for hijackers, terrorists, thieves, and saboteurs; they need very special protection. The more nuclear transports there are, the greater the risk of interference. The societal consequences of nuclear transportation, like those of the protection of commercial reprocessing plants, will be considerable.

Uranium enriched with the rare isotope uranium–235 can also be used to construct nuclear weapons. The technology for enriching uranium, in which the percentage of uranium–235 in natural uranium is increased, is now well known. Many countries could build a small uranium-enrichment plant. Pakistan is in the process of doing so and South Africa has already done so.

Conclusions

Although the most sophisticated weapons get into the arsenals of many Third World countries, only the industrialized countries will be able to afford the wide range of surveillance and target-acquisition systems and automated C3I systems, in other words, the military superstructure, needed for the fully automated battlefield.

But while the industrialized countries are acquiring their automated systems, the less developed countries will acquire, mainly through the global arms trade, increasingly accurate weapons of great fire-power. This is bound to fuel regional arms races and increase regional instabilities.

Very destructive conventional weapons, and very accurate delivery systems for them, will also be acquired by sub-national groups. And, as the capability to produce nuclear weapons spreads worldwide because of the spread of peaceful nuclear technologies, terrorist groups will also acquire the capability to make nuclear weapons.

Terrorists do not need sophisticated delivery systems for nuclear weapons. They could deliver these weapons in suitcases or explode them from hidden locations in buildings or vehicles. But some of the missiles that will become available to sub-national groups could carry warheads containing weapons of mass destruction.

The threats from sub-national groups will become more serious and harder to contain as urban guerrilla violence increases because of larger world populations and more urbanization. And terrorist groups are likely to combine to increase their power. These trends will increase regional instabilities and the danger that violence begun by sub-national groups will escalate to inter-state war.

Chapter 9
The Consequences of Automated Warfare

Technologically, fully automated warfare is foreseeable within the next twenty or thirty years. Unless, that is, military science and technology are brought under control in the meantime.

As we have seen, in its simplest form an automated battle would take place in an area of territory from which all people had been evacuated. One side would attack the area with unmanned vehicles; the other side would defend it with autonomous missiles. The entire process could be automated. If there was any human control, it would be remote; human operators would stay well away from the battlefield.

Will the new technologies work? Experience shows that complex technologies do work. Modern cars, television sets, electronic calculators, and so on, rarely break down even though they are much more complex than earlier models. And we happily assume that jumbo jets, the computers that our banks use, the computers that diagnose disease, etc., will not let us down.

It is true that some relatively expensive sensors used in missiles, including infra-red and radar sensors, can be deceived by rather simple and cheap countermeasures like flares and chaff. One commentator tells the story of 'a costly piece of automated radar-linked air defense equipment which, on a demonstration, instead of shooting down a target aircraft, blew up a nearby field latrine with a ventilator which had a similar effect on the radar as the aircraft's turbofan engine'.

Whether or not missiles, for example, operate effectively will depend on whether enough use is made of the signals coming from the battlefield. If the missile is provided with enough sensors to make use of all, or many, of the signals, then it will correctly interpret the scene and perform its task efficiently. Also, the designer knows how the system will perform under various conditions and can accurately describe its capabilities to the user.

The main problem with automated weapon systems is, in fact, not technological but organizational. There are a large number of sub-systems involved. Putting them together to perform specific tasks is like completing a complex jig-saw puzzle.

The industrialized countries are the first to acquire a range of automated military systems. The weapons for the automated battlefield, and their supporting technologies, will almost certainly first come to Europe where the world's two major military alliances stand eyeball to eyeball. That they will eventually transform warfare is beyond doubt.

Discussions are now going on in NATO countries about how new technologies should be used in the interim period. This debate will be described in the next chapter. In this chapter we will discuss the consequences of automated warfare.

The momentum of technology

The automation of warfare is occurring for a number of reasons. One is simply that the technologies are becoming available and every conceivable technological advance is, sooner or later, used for military purposes. A combination of pressures – from the military, the weapons industry, the bureaucracies that administer the military establishments and the military scientists and technologists – in favour of this military use of technology has so far proved well-nigh irresistible, politically.

Political leaders normally give in to this pressure and agree to the deployment of virtually all the weapons and weapon systems the military scientists develop. This will go on unless and until sufficiently strong pressure from public opinion against continuous militarization is exerted on the political leaders to overcome the pressure exerted in favour of it.

The USA is particularly keen on maximizing its use of military technology and is, therefore, at the forefront of developments leading to the automation of warfare. The Americans are making a great effort to retain their technological superiority in as many weapon systems as possible. The Soviets, in turn, are trying to eliminate the American superiority and catch up.

A basic American belief is that, provided enough money is given to scientists for research, the nature of their society, particularly its openness, is such as to encourage innovation more than the Soviet system does. There may be some truth in this. History shows that the more open a society is, the more able it is to maximize the skills available to it.

In weapon development, money can be a constraint. But the main constraint is the skills that can be mobilized. The more efficiently these skills are used, the more rapid is the development. And here, generally speaking, the Americans, and the West as a whole, have the advantage over the Soviets. This is dramatically shown by the fact that in the USA about 125,000 research scientists and engineers are in military research and development. The corresponding number in the USSR is 250,000. And yet the Americans are ahead in military science and technology. American military scientists work more than twice as effectively as their Soviet counterparts.

Weaponry is a field of activity, and perhaps the only one, in which the USA hopes to be able to keep permanently ahead of its Soviet rival. To fall behind would, according to many Americans, bring political and military disaster, enabling the Soviet Union to spread its influence and ideology worldwide. The Americans seem to believe this almost as an article of faith.

The struggle by the USA to retain its military pre-eminence and by the USSR to catch up is the main force driving the on-going East-West arms race in both nuclear and conventional weapons. This race will go on until both sides stop attaching supreme importance to technological advantage.

In the meantime, the momentum of military technology will inevitably lead to the increasing automation of warfare. Here, the key technology is data-processing in which the West has a commanding lead and is likely to keep it, and probably widen it, for the foreseeable future. It is because of this lead that automated warfare systems are first being deployed by NATO's armed forces.

Manpower shortages and the increasing lethality of the battlefield are other reasons for the development of the automated battlefield. In the USA, for example, the male population between the ages of 17 and 19, the ages of greatest interest to military recruiters, will decrease sharply. By the year 2000, the number of American males in this age group will have dropped by 27 per cent, compared with the current figure.

Moreover, as time goes on and weapon systems become more complex, those in the armed services must have more sophisticated skills. More time and resources are, therefore, devoted to military training and soldiers become too valuable to lose in large numbers. Mass-produced robots will be cheaper and more expendable than humans.

The concept of automated warfare raises basic questions. How would victory be defined in an automated battle? Is it necessary for blood to be spilled in war? If warfare is becoming a battle between unmanned vehicles and robots, on the one hand, and automated missiles on the other, why not simply decide the issue by having the generals play computer games? In fact, isn't the concept of automated warfare so absurd as to make a conventional war in Europe incredible?

The answers to these questions depend on the reasons why countries go to war and on how ideas about the meaning of 'national security' evolve during the period in which automated weapons are being developed and deployed.

Victory in an automated battle may well go to the side that can keep up the battle for the longest time. In an automated battle, weapons would be used up at a very fast rate. This would put a high premium on establishing the large-scale production of unmanned vehicles, smart missiles and other automated weapon systems.

Industries to produce these weapons and weapon systems would, of course, have to be set up long before the battle began. Automated warfare could, therefore, lead to the militarization of the economy and society. That this development is probable is shown by the experience of Israel which uses the most advanced technologies for military purposes because of its shortage of manpower and the resources to acquire very large numbers of weapons. But the price Israel pays for its high-quality technological arsenal is the need to keep its economy permanently on a war footing and all the disadvantages that go with a militarized society.

Changing attitudes to security

Margaret Mead, the anthropologist, stated in 1968: 'Warfare exists if the conflict is organized and socially sanctioned, and the killing is not regarded as murder.' Her criteria for war are 'organization for the purpose of a combat involving the intention to kill and the willingness to die, social sanction for this behaviour, which distinguishes it from murder of members of its own group, and the agreement between the groups involved on the legitimacy of the fighting and the intent to kill'.

Most people would have accepted Margaret Mead's definition when she formulated it. And they would have argued that countries would have gone to war, often as a last resort, to defend their territory from invasion and occupation by other countries anxious to increase their territory.

In the past twenty or so years, however, ideas about security, war and peace have changed. For one thing, the spectrum of warfare has broadened, from terrorism and unconventional warfare at one end to strategic nuclear world war at the other. For another thing, the definition of security has become much broader.

National security used to be limited to defending one's territory against an external military threat. The defence of the 'political and social values' of the society was then added. Threats to these values were first seen to be external threats, but they came to include perceived internal threats. This aspect of national security became important for Western countries, for example, after the 1917 Russian revolution. A third element of national security is the need to defend sources of strategic raw materials and strategic markets. In particular, NATO European countries and Japan are concerned to defend their supplies of oil from the Middle East. The currently stated missions of NATO, 'to defend the territorial integrity of NATO and to defeat the Warsaw Pact, to deny the Soviets the oil resources of the Middle East and to maintain a strategic reserve in the Western Hemisphere', reflect these three elements of national security.

In today's world, there are also non-military threats to national security and some of these threats may be, or may become, more serious than military threats.

The rapidly increasing world population is one non-military threat to security. For every two people now on Earth there are likely to be five in the year 2030, less than fifty years from now. This large increase will take place mostly in the Third World. The population of the industrialized countries, about 1,200 million, will remain roughly constant. The Third World population, however, is likely to reach 10,000 million.

These dramatic population changes will have far-reaching security consequences. The USSR, for example, with a more or less constant population of about 200 million people, has China just to the east of it and India to the south. The populations of these countries, already about 2,000 million, will increase significantly. These statistics must affect Moscow's thinking about Soviet security as much as, if not more than, relations with NATO countries.

Another non-military threat to national security is the threat to the environment. We have heard much recently about acid rain that is destroying the forests of West Germany, the lakes and forests of Scandinavia and Canada, and so on. The forests and lakes are valuable national resources. When they are destroyed by the actions of neighbours, even though these countries may be allies in the military sense, the issue becomes one of security.

Then there is the carbon dioxide problem. The burning of ever greater amounts of fossil fuels is significantly increasing the amount of sulphur dioxide in the atmosphere (the substance that makes the rain acidic), and it is also increasing the amount of carbon dioxide in the atmosphere. More carbon dioxide in the atmosphere in turn increases the amount of ultra-violet light that enters the atmosphere and the global temperature increases – the so-called 'greenhouse effect'. Because trees remove carbon dioxide from the atmosphere, the destruction of the world's rain forests aggravates the problem.

The heating of the atmosphere will have far-reaching consequences for the world's food production. But it will not affect all regions of the world equally. North American grain production will probably suffer whereas Soviet grain production will benefit, as will the main rice producers. Food supplies are an important element of security. Changes in the amounts of food produced may change regional and global balances of power.

The economic performance of a country is, we now realize, an important national security issue. Low economic growth, high inflation and high unemployment are often as serious, if not more serious, threats to political and social values, and hence to security, than any external military threats.

The widening poverty gap between the rich and the poor countries is increasing North–South tension to such an extent that it may soon exceed East–West tension. Many people, therefore, believe that economic development and the evolution of a new international economic order are important for the security of all.

An important element of security is non-renewable resources. The only significant security issue in recent years between, for example, the UK and Denmark has been the right to harvest dwindling stocks of fish in the North Sea. This led to the 'cod wars' between the two countries. In another example, Israel is, to say the least, reluctant to give up the water table on the Palestinian West Bank for fear of jeopardizing the country's agricultural crops, an important resource. This factor may well be the one responsible for the past forty years of instability in the Middle East. Much of today's unrest in Central America is caused by a shortage of arable land. As populations increase in the region, the demand for land increases, but land-owners are not prepared to give up land without a struggle.

Measures taken to counter perceived military threats to security often make the non-military threats worse. For example, the adverse consequences of increasing populations, North–

South tension and shortages of non-renewable resources could be alleviated if the rich countries would put resources into Third World development to improve education, health and general prosperity.

Money is important to solve major Third World problems, but scientific and technically skilled manpower is even more important. The financial and manpower resources needed for development are increasingly tied up by the military. Many of the scientists best able to solve Third World problems of energy production, food production, health, and so on are now spending all their time developing weapons.

How conflicts may escalate

Reducing conflict in the Third World is important because the most likely way in which a nuclear world war will occur is through a conflict in a Third World region. Such a conflict – in, say, the Middle East, the Persian Gulf, southern Africa, and Korea – may begin as a conventional war and escalate to a local nuclear war in which nuclear weapons produced by the local powers are used. This war may then spread to Europe, begin there as a conventional war and then escalate to an all-out nuclear war between the USA and the USSR.

The superpowers become involved in wars in the Third World mainly because they provide most of the weapons with which these wars are fought. A feature of modern war is that munitions are used up at a very rapid rate. And, as war becomes more automated, this rate, particularly for missiles, increases dramatically.

Consequently, countries at war need to replenish their arsenals frequently. This was shown dramatically during the 1973 war in the Middle East. Within a few days of the outbreak of the war, the superpowers had to airlift munitions to their respective clients to prevent their defeat.

The superpowers cannot afford to let their clients be beaten in war because they would lose their credibility with their other allies. But the need for each superpower to guarantee the survival of its clients can easily lead to a clash between the superpowers. And this is why the arms trade is a major link between war in the Third World and a nuclear world war.

Military technological advances, particularly the development of automated nuclear-weapon systems, also make more likely the escalation of a war in the Third World to a nuclear world war. New nuclear weapons, for example, are so accurate and reliable that they are seen as more suitable for fighting a nuclear war than deterring one.

Evolving nuclear policies

Using modern guidance systems, intercontinental ballistic missiles can be aimed with great accuracy at small military targets 10,000 kilometres away. Soon, submarine-launched ballistic missiles, like the American Trident–II, will have similar accuracies.

The deployment of these weapons, together with very accurate tactical nuclear weapons, is causing the superpowers to change their nuclear policies from nuclear deterrence by mutual assured destruction to nuclear-war fighting. The reason for this change is that one cannot operate a policy of nuclear deterrence by mutual assured destruction with very accurate nuclear weapons. The policy depends on the theory that if the enemy knows that if he attacks you then his cities and industries will be destroyed in retaliation he will not attack you in the first place. The cities are, in other words, the hostages to nuclear deterrence.

But as you increase the accuracy of your missiles you are more and more able to target the enemy's military forces. Eventually, you will be able to target his strategic nuclear forces with your nuclear missiles. The enemy will then believe that you are targeting his nuclear forces and not his cities and the hostages to your deterrence disappears. At this point, you will most likely change your policy to nuclear-war fighting in which military targets are given a higher priority for attack.

The more the superpowers adapt their military tactics and

strategies to nuclear-war fighting the greater will be the probability of nuclear war, because the idea that a nuclear war is fightable and winnable will gain ground. When a nuclear-war fighting policy is adopted the political control over the use of nuclear weapons weakens and military control increases. This is particularly true if large numbers of tactical nuclear-war fighting weapons are deployed because they will become integrated into military tactics. Once a war begins, military tactics cannot be radically changed and, therefore, permission to use nuclear weapons has to be given to relatively junior military commanders.

As nuclear war-fighting weapons are being deployed, military scientists in both superpowers are energetically developing technologies to support them. The most important of these are anti-submarine warfare technologies, anti-ballistic missile systems and anti-satellite systems. Of key importance to these technologies is the ability to collect and rapidly process large amounts of data. As computer power increases, therefore, the technologies will become more effective.

When one side perceives that its anti-submarine warfare, anti-ballistic missile and anti-satellite systems are effective it is likely to change its nuclear policy again, from nuclear war-fighting to nuclear war-winning. This change will occur because of the perception that a considerable advantage can be obtained by making a pre-emptive nuclear attack on the other side.

Currently, land-based intercontinental ballistic missiles are vulnerable to a surprise attack. Each side has accurate enough nuclear weapons to destroy the other side's land-based missiles in their hardened silos. But strategic nuclear submarines are not yet so vulnerable. Neither side can detect and destroy at the same time all the strategic nuclear submarines the other side has at sea. As anti-submarine warfare techniques improve, however, the time will come when one side perceives that it can do so.

In the race for effective anti-submarine warfare the Americans have considerable advantages, not only because they are ahead of the Soviets in the relevant technologies. Although the Soviets have more strategic nuclear submarines – they have 62 whereas the Americans have 32 – they are able to keep a much smaller number at sea at any one time – about 6 compared with over 20 – because they have an insufficient number of trained submariners and are unable to afford the costs of maintaining a large submarine fleet. The task for American anti-submarine warfare systems is, therefore, much easier than that for the Soviet ones. Also, Soviet submarines are boxed in. They have to go through relatively narrow channels (like those between the north of Scotland and Iceland and between Iceland and the Arctic ice) to get from their home ports to both the Atlantic and Pacific Oceans. These channels can be, and are being, effectively monitored by anti-submarine warfare sensors.

No military planner, even the most optimistic, would assume that he could destroy all the enemy's land-based and sea-based nuclear forces in a sudden attack out of the blue. Some will escape. But those that do could be destroyed by anti-ballistic missiles. If one side wanted to make a first strike against the other side, the first thing it would want to do would be to destroy the enemy's satellites in their orbits, his 'eyes and ears' in space. Hence the importance placed by the superpowers on developing anti-ballistic missile and anti-satellite warfare systems.

Moves to a nuclear war-winning, or first-strike, capability will again increase the risk of a nuclear world war. Nevertheless, the new military technologies, when applied to nuclear weapons and their supporting technologies, will cause nuclear policies to shift from nuclear war-fighting to nuclear war-winning.

Fears about the survival of nuclear forces in a pre-emptive attack may well provoke the superpowers to deploy launch-on-warning systems. In such a system, the early-warning-satellites of one side would, if they detected the launching of enemy ballistic missiles, send signals directly to the computers controlling the firing of that side's ballistic missiles. The computers would automatically and instantly fire the missiles so that they would be gone when the enemy missiles arrived. The missiles would destroy empty silos. Launch-on-warning is the ultimate in automated warfare. To leave the decision to launch the nuclear holocaust to a computer is also, many would agree, the ultimate in human madness.

Many of the threats to the security of humankind outlined

above are mutually reinforcing. If there is a nuclear catastrophe, it is most likely to come about because of a combination of several factors. Competition for raw materials is likely to be very intense when a superpower may perceive that it has a nuclear first-strike capability. Combine these elements with a war in a Third World region where superpower rivalry is high, irresponsible leadership in one of the superpowers, and high tension in Europe (because, say, of serious unrest in East Germany and armed groups in West Germany threatening to intervene in East Germany), and you have a plausible recipe for world disaster.

The danger of the escalation of violence in the Third World to a nuclear catastrophe is heightened by the sheer number of conflicts in the Third World. Since the Second World War, there have been about 200 wars, almost all of them in the Third World. These wars have taken place in all the continents except Australasia and have involved the armed forces of about eighty-five countries. About 20 million people have been killed in them. At any one time, there are, on average, twelve wars going on in the world, and a new war begins in the Third World, on average, every three months.

To this high level of state violence must be added increasing sub-national violence, particularly terrorist activity. All these conflicts, national and sub-national, are being fought with weapons of increasing fire-power and accuracy and with increasingly automated weapon systems. It will be more and more difficult to stop violence from escalating to wars across borders, threatening world security.

Moves to automated warfare in Europe will probably further destabilize the international situation. Because it will make conventional tactical warfare in Europe incredible, it is likely to increase the risk of strategic nuclear war. Any conflict in Europe is likely almost immediately to escalate to the strategic nuclear level.

Countries geared up for automated warfare are unlikely to be able to deal effectively with conflicts fought with weapons and weapon systems that are much less automated. Consequently, the main danger of automated warfare is that it will increase the importance of both ends of the spectrum of violence – of terrorism and unconventional warfare, and of strategic nuclear war – and make more likely the rapid escalation of sub-national violence to all-out nuclear war.

Chapter 10

Offence or Defence?

In the words of Gary Chapman, Executive Director of Computer Professionals for Social Responsibility:

> There is a massive trend towards the militarization of computer science in the United States. The Department of Defense is engaged in a large research and development program for putting advanced computer technology into conventional weapon systems. Computer science is on the threshold of a much higher level of capability – one that ostensibly will allow computers to 'think' and reason in ways that resemble humans. The work that has led to this development is broadly categorized as 'artifical intelligence'. In the United States, nearly all research and development funding for artificial intelligence has come from a single agency within the Department of Defense, the Defense Advanced Research Projects Agency (DARPA). Recent comments from DARPA staffers now indicate that the Department of Defense wants to cash in the chips it has acquired in advanced computer science by demanding real, operational weapons development.

DARPA is well into its $600 million five-year Strategic Computing Initiative to build an automated tank, automated devices for flying aircraft (like the stealth aircraft which will be flown wholly by computer), automated battle management systems, and so on. The Pentagon's goal is to give machine intelligence the job of waging war without human intervention. If the enemy uses human soldiers on the battlefield they will be killed by computerized weapons. If the enemy uses robots and automated machines on the battlefield they will be destroyed by computerized weapons. As we have stressed in this book, there is no technological reason why the Pentagon should not achieve its goal of the automated battlefield.

As we have seen, the battlefield is becoming too dangerous for humans because very lethal conventional weapons can be delivered on targets with great accuracy. Also, in the USA, for example, by the year 2000 only a minority of the population will be under forty years of age. The group from which the military recruit is shrinking fast. Automated weapons are needed to replace soldiers.

The military do not, however, intend to wait until the battlefield can be completely automated before deploying computerized weapons. *The battlefield is being automated in stages.* NATO military officers, for example, are now considering using autonomous weapon systems to locate, identify and attack targets deep behind enemy lines. These new battle plans involve the use of what the military call 'emerging technologies' or ET weapon systems.

ET weapons will react very rapidly. Surveillance systems will detect enemy forces on the move, at great distances, and guide missiles to attack them so quickly that there will be simply no time for humans to intervene. Human judgement will be irrelevant. Commanders of ET forces will be mere onlookers from a great distance, watching their computerized attacks in comfort on television.

The replacement of soldiers by military machines may appear attractive to some, but there are grave problems when automated weapons are used at long range for offensive purposes. As Gary Chapman points out, one problem is that military computer programmers will, to say the least, find it extremely difficult to distinguish combatants from non-combatants, a distinction that is 'at the heart of all rules of the conduct of war'. To cross the threshold of 'allowing machines to kill humans with nonchalance and without regret' is to move into an age of 'new barbarism'.

A less objectional use of the new weapons is to rely on short-range weapons and to change military doctrines to emphasize defence rather than offence. As will be shown in this chapter, a conventional defensive deterrent is the most cost-effective way of providing real national security. It would also move us away from the 'new barbarism'.

Currently, the debate about the best use of emerging technologies mainly relates to the need to solve fundamental problems with NATO's military policy, although the issue has relevance to most countries' security problems. To put the topic into perspective it is useful to describe briefly what is wrong with NATO's military posture.

NATO's weaknesses

Until the early 1960s, NATO's strategy was massive relatiation. If the Warsaw Pact attacked Western Europe, NATO would respond by dropping nuclear weapons on Soviet cities. But in the early 1960s NATO policy changed to flexible response.

The reason for this change was simple. The Soviet nuclear arsenal had grown sufficiently to make NATO's strategy of massive retaliation no longer tenable. It was all very well to threaten the Soviets with genocide when they could not reply in kind. But when they could, NATO had to change its tune. And so NATO took on flexible response. The idea of flexible response is that if the Warsaw Pact should attack NATO we will try to beat off the attack and win the war with conventional weapons. But, if we could not, we would be prepared to escalate the war and use small nuclear weapons. The other side, it is hoped, would play the game and respond with only small nuclear weapons. We would then try to negotiate with the Soviets and stop the war. If the negotiations failed we would go to bigger nuclear weapons, to the next rung of the escalation ladder. We would fight the nuclear war at this level and negotiate again. And so on. The final step in this ladder of escalation is the use of strategic nuclear weapons. Genocide is not abandoned, just delayed. And, of course, now that the Soviet Union has achieved a balance of terror, genocide would be mutual. This is why the policy is called nuclear deterrence by mutual assured destruction.

Morton Halperin, when he was US Deputy Assistant Secretary of Defense, described NATO's policy of flexible response and mutual assured destruction very aptly. 'The NATO doctrine is,' he said, 'that we will fight with conventional forces until we are losing, then we will fight with tactical nuclear weapons until we are losing, and then we will blow up the world.'

Flexible response, however much the authorities deny it, is a nuclear war-fighting strategy. The official way of putting it is to say that Nato would aim to 'prevail', which means win, at all levels of conflict, conventional and nuclear, and be prepared to fight a 'prolonged nuclear war'.

But NATO's policy is simply no longer credible. Any use of nuclear weapons would almost certainly rapidly escalate and destroy Europe, not to speak of the USA.

And so no sane political leader would, under any circumstances, agree to the use of nuclear weapons. Nevertheless, it is still the official NATO policy. Caspar Weinberger, the US Secretary for Defense, said so emphatically in his 1985 Annual Report to Congress. 'Flexible response continues to be our policy,' he said, 'and will remain so throughout the President's second term.'

The incredibility of NATO's policy, based as it is on the early first use of nuclear weapons, has presented the military with a dilemma. The conventional NATO wisdom is that if Warsaw Pact forces attacked Western Europe they would probably win the war in a few weeks, or even days, unless NATO used tactical nuclear weapons to stop the advancing forces. General Bernard Rogers, NATO's top military commander, put it this way in a recent speech to the Dutch Parliament:

> NATO's current military posture will require us – if attacked conventionally – to escalate fairly quickly to the second response of our strategy, 'deliberate escalation' to nuclear weapons. The plain fact is that we have built ourselves a short war; we simply are unable to sustain ourselves for long with manpower, ammunition, and war reserve stocks to

> replace battlefield losses and expenditures. Therefore, we face the serious risk of having no recourse other than the use of nuclear weapons to defend our soil.

But as General Rogers himself said, this use of nuclear weapons would, in all probability, escalate.

> I do not think that a limited nuclear war in Europe is possible. No, if we have to resort to the use of nuclear weapons under current conditions if attacked or if . . . they [the Warsaw Pact] resort to the use of nuclear weapons, in my opinion it cannot be limited to Western Europe. I happen to be one who believes there would be fairly quick escalation to the strategic level. I really think it would escalate and could not be limited.

General Rogers' solution to NATO's dilemma of having to choose between nuclear suicide and capitulation is that NATO should use new weapons and military technologies to attack Warsaw Pact forces being brought up to the front line at the same time as NATO is attacking the invading Warsaw Pact forces at the front line. The General argues that if he could stop the flow of Warsaw Pact reinforcements, the momentum of the attack would be broken and it would fail. Specifically, General Rogers wants to use for this purpose real-time surveillance and target-acquisition systems and very powerful conventional warheads, accurately delivered by long-range ballistic missiles, on targets far behind the front line. The idea is not to do away with all NATO nuclear weapons. The General wants to keep some nuclear weapons in Europe. He argues that the presence of tactical nuclear weapons prevents the other side from concentrating their forces, to avoid giving NATO good targets for its nuclear weapons. And if the Warsaw Pact forces do not concentrate, they are unable to make an effective massive conventional attack.

NATO also believes that its threat to use nuclear weapons, even though such use would be suicidal, in some way deters the Soviets from starting both a conventional and nuclear war. Or, the implication is that the Soviets would be so uncertain about whether or not NATO would use nuclear weapons if facing defeat in a conventional battle that they would not risk an attack in the first place. Many believe, though, that the argument that uncertainty is an adequate deterrent is unreasonably flimsy.

The Rogers plan would reduce reliance on short-range nuclear weapons, the so-called battlefield nuclear weapons, while keeping medium- and long-range nuclear weapons in Europe. Battlefield nuclear weapons are those with ranges of less than about thirty kilometres. These include nuclear artillery shells, ground-to-air (air defence) missiles, atomic demolition munitions (nuclear land mines) and gravity bombs to be dropped on the battlefield by ground-support aircraft. Of the 6,000 nuclear weapons now deployed in NATO West European countries, about 4,000 are battlefield nuclear weapons. The rest are nuclear warheads for Lance and Pershing-II ground-to-ground missiles, ground-launched cruise missiles and aircraft bombs.

There is no doubt that battlefield nuclear weapons are, to say the least, extremely undesirable weapons. Because they have short ranges many are kept close to the East–West German border. They, would, therefore, have to be used very early in a war or they might be captured; they are 'use them or lose them' weapons. Battlefield nuclear weapons would ensure that a war in Europe would very rapidly escalate from conventional to nuclear war. The main military targets for battlefield nuclear weapons – enemy tanks and aircraft – can now be attacked very effectively with conventional weapons. So there are no reasons for having nuclear weapons for these purposes.

General Rogers' plan, also called Follow-On Forces Attack or FOFA, calls for much deeper attacks in enemy territory than NATO can make today. Currently, NATO's capability to attack targets in Warsaw Pact territory with conventional weapons is limited to attacks by manned aircraft – US F–111 aircraft and British Buccaneers and Tornadoes. As Warsaw Pact defences improve, the distance these expensive aircraft can effectively penetrate into enemy territory decreases. NATO is unable to attack targets more than about fifty kilometres behind the lines. Greater penetration would involve unacceptable losses to aircraft. Whereas NATO is currently limited to attacking targets in Eastern Europe within fifty kilometres or so of the East–West border, the FOFA plan involves attacking targets with ballistic missiles at distances of up to 300 or so kilometres.

The Rogers plan, if it became NATO policy, would almost certainly provoke the Soviets to deploy similar offensive long-range conventional weapons and would, therefore, trigger off a conventional arms race in Europe. Many believe that this would become as destabilizing for Europe as is the current nuclear arms race. This 'arms race' instability is perhaps the most serious objection to FOFA.

But it is not the only serious objection to the plan. Another is that it would introduce a new element of 'crisis instability' into European affairs. The deployment of many long-range ballistic missiles carrying highly destructive conventional warheads and the capability of very accurately delivering these warheads would increase Soviet perceptions of possible aggressive actions by NATO. NATO would seem to the other side to be a much more aggressive alliance than it is now, particularly during times of crisis. This could greatly increase the probability that the Warsaw Pact would make a pre-emptive attack at a time of crisis, thereby increasing the risk of unintentional war.

Implementation of FOFA would also introduce new 'arms control' instabilities. The deployment of weapons and their supporting technologies able to strike deep in Warsaw Pact territory would almost certainly worsen relations between the Soviet and American leaders. The distrust between the two sides would deepen and deteriorating relations become very difficult to repair. The atmosphere for arms control would deteriorate. Arms control treaties would then become even more difficult to negotiate than they are now.

The objections to the FOFA concept so far described – that it would increase crisis instability in Europe, stimulate a new conventional arms race and worsen the atmosphere for arms control – are political. There are also a number of military objections.

It is not certain whether the latest Soviet military tactics depend on reinforcements brought up from distant rear areas. The tasks assumed by FOFA for long-range conventional forces may, therefore, not exist. And, if they do exist, the Warsaw Pact may well change its tactics to ensure that its reinforcements do not present targets for new NATO conventional weapons.

It should be explained that, although the General's opinion is the conventional wisdom in NATO headquarters, there is a sizeable minority of NATO military men who do not agree that NATO's conventional forces are as weak, relative to those of the Warsaw Pact, as General Rogers would have us believe. These officers explain that Warsaw Pact forces also have serious weaknesses, particularly in logistics and C3I.

It is, therefore, reasonable to question General Rogers' basic concern that the Warsaw Pact could make a surprise attack with massed armour and artillery across, say, the North German plains, rapidly rupture NATO's defences, burst out into West Germany and the Benelux countries, and disrupt the political and social systems of these countries. One can, for example, question whether the Warsaw Pact has enough *reliable* combat troops for a rapid breakthrough into West Germany. (Could the Soviets rely on Polish troops to fight effectively in another European war, could they rely on Czechoslovakian troops, and so on?)

Also, even if the Warsaw Pact has enough reliable combat troops, could it concentrate the large forces needed for a breakthrough, given its weaknesses in logistics and C3I? And, if the Warsaw Pact could concentrate enough troops for a breakthrough, could it take NATO by surprise? Given the effectiveness of modern intelligence systems, such as reconnaissance satellites, it seems, to say the least, doubtful.

In addition to the political and military objections to FOFA there are economic ones. Implementation of FOFA would require higher military budgets. Given the economic situation in most NATO European countries, there is little prospect that extra money will be made available for FOFA. European populations could not be easily persuaded to foot the FOFA bill, mainly because of the instabilities it would introduce. Western Europeans are likely to be persuaded to spend more money on the military in the current period of low economic growth only for military policies that demonstrably increase stability and security in Europe, and decrease the risk of war in Europe.

FOFA depends largely on weapons, mainly long-range ballistic missiles, and supporting technologies that would be mainly supplied by American industry. European defence industries do not generally produce long-range ballistic missiles, but they do produce effective short-range missiles, particularly anti-tank, anti-aircraft and anti-ship missile systems, and would, therefore, prefer military postures that use these weapons. These industries, naturally anxious to get their share of the military cake, are not in favour of the Rogers plan.

We can conclude that an acceptable solution to NATO's problems should satisfy the following six criteria:

it should provide an effective defence;
it should increase 'crisis stability';
it should not provoke a new arms race;
it should improve the climate for arms control;
it should be acceptable to the public;
it should be affordable.

FOFA does not satisfy any of these criteria.

Defensive deterrence

As we have seen in earlier chapters, new military technologies increasingly favour defence over offence. Defence is becoming increasingly cost-effective in that it is cheaper to destroy weapon systems like main battle tanks, long-range combat aircraft and large warships than to deploy them. The shortcomings of FOFA, and other deep-strike concepts, and the increasing cost-effectiveness of defence have raised considerable interest in alternative military defence policies for European NATO countries based on purely defensive conventional deterrence.

One proposal for defensive deterrence in Europe is based on the concept of 'non-provocative' defence. This relies on the principle that the size, structure, weapons, logistics, training, manoeuvres, war games, military-academy textbooks, and other activities of the military forces of a country can be so designed as to demonstrate *in their totality* that they provide an effective conventional defence but have virtually no offensive capability and that they do not rely on tactical nuclear weapons. The military forces could, on request, be opened for inspection by neighbouring, or any other, countries to assure them of the non-aggressive, non-threatening nature of the forces.

In Western Europe, a non-provocative, non-nuclear defence would look something like the following. A defence zone, some 50 kilometres deep, would be prepared all along the 1,000-kilometre East–West border. This defence zone would be saturated with all kinds of ground-based sensors, and provided with a vast network of underground fibre-glass cables for relatively invulnerable communications and with numerous positions for troops to take cover.

There would be border surveillance by, for example, remotely piloted vehicles, airborne side-looking radars and electro-optical sensors, other airborne sensors and, if available, reconnaissance satellites. Airborne side-looking radar and electro-optical systems would also be used for target acquisition in real time.

Many anti-tank obstacles would be built in the border area and much use would be made of smart anti-tank mines (i.e. mines able to identify tanks and destroy them). In particular, an extensive network of 'explosive hoses' would be buried along the border. Liquid explosives can be pumped through these plastic tubes and exploded when tanks pass over them. Equipment capable of moving large amounts of earth would be available to produce anti-tank obstacles. Preparations would be made to fire anti-tank mines with artillery and multi-role rocket systems into the defence zone. The purpose of the anti-tank obstacles and mine-fields would be to delay and stop enemy tanks so that they could be destroyed by missile fire and also to channel tanks into areas where they could be bombarded and destroyed.

Details of the defence zone would be fed into computers, so that any hostile troop or tank movements could be located and identified very accurately and destroyed by pre-planned fire using, for example, an artillery-targeting grid prepared in peacetime. Very mobile squads of troops would be available to operate in the area. They would be trained in the area so that they would become very familiar with it.

The troops would be armed with a judicious mixture of effective anti-tank missiles and anti-tank cannons, anti-aircraft missiles and light anti-aircraft guns. Emphasis would be given to simply-operated and expendable missiles, cheap to produce in large quantities.

Command, control, communications and intelligence systems would be mainly automated but decentralized. Highly mobile squads armed with weapons of high fire-power, mainly missiles, would also be deployed to deal with any enemy forces that broke through the forward defence zone. Troops would be dispersed throughout the country to deter attacks by airborne forces and to defend coastal areas.

The armed forces would not have main battle tanks, long-range combat aircraft or large warships. Nor would they have long-range airlift capability. The ranges of missiles would be no more than those required to bombard the defence zone, so that they would be non-provocative. Maximum missile ranges would be roughly eighty kilometres. Armoured vehicles would be limited to very mobile, short-range and relatively lightly armoured vehicles. Large armoured units would give way to small, light mechanized infantry units. Combat aircraft would be limited to single-role interceptors and close-support ground-attack aircraft. Much use would be made of helicopters for a variety of tasks, including anti-tank warfare, ferrying troops, and so on. Naval forces would rely on missile-armed fast patrol boats, equipped with anti-ship missiles, and small diesel-powered submarines.

Complex technology would never be used for its own sake, but weapons would be chosen for specific tasks and operated well within their design characteristics. Most important, the defence would be decentralized and would, therefore, not provide targets for enemy bombardment.

Advocates of non-provocative conventional defence for NATO European countries claim that it would overcome many of NATO's disadvantages and have many potential advantages. It would provide a credible defence for Europe without reliance on conventional offensive weapons or tactical nuclear weapons.

If non-provocative defence is adopted, therefore, tactical nuclear weapons can be removed from the European land-mass.

The removal of these nuclear weapons would considerably improve European security. If they are left in Europe, they will have to be replaced because nuclear weapons have a 'shelf-life' of about twenty-five or thirty years. The new nuclear weapons would be much more accurate, and, therefore, seen by the military to be more useful for fighting rather than deterring a nuclear war. If significant numbers of these new weapons are deployed, the military will come to believe in the 'fightability and winnability' of nuclear war and the risk of such a war will sharply increase.

The best way of reducing military reliance on tactical nuclear weapons would be to adopt a no-first-use policy in which it is declared that under no circumstances would NATO be the first to use nuclear weapons. A no-first-use policy would recognize the fact that the *only* function of nuclear weapons is to deter the other side from using its strategic nuclear weapons. In other words, one side's nuclear weapons are there to 'neutralize' those of the other side. Nuclear weapons are militarily unusable weapons.

Once this fact of life in the nuclear age is recognized, tactical nuclear weapons are seen to be redundant. Nuclear deterrence should be left totally to strategic nuclear weapons. The only purpose of these nuclear weapons would be to deter the other side from using their nuclear weapons, they would never be used in a first strike.

The main advantage of a no-first-use policy is that it would convince the military that they would not be given permission by NATO political leaders to use tactical nuclear weapons and would, therefore, not integrate them into their tactics. The military would not then come to believe that nuclear wars are fightable and winnable.

An effective conventional defensive deterrent would allow the adoption of a no-first-use policy by NATO and make possible the removal of all tactical nuclear weapons from the European land-mass. To the extent that a nuclear deterrent is felt to be necessary, it can best be provided by submarine-launched ballistic missiles deployed on strategic nuclear submarines.

Because defensive weapons, like anti-tank, anti-aircraft, anti-ship missiles and so on, are more cost-effective than main battle tanks, long-range aircraft and large warships, necessary weapons for the invasion and occupation of a country, an effective non-provocative defence would be relatively cheap. The fact is that if a country wants to use its military money to defend its territory against invasion and occupation in the most effective way it would invest in these defensive weapons.

A conventional defensive deterrent in Europe would remove one cause of the East–West arms race – one side reacting to the threat it perceives from the deployment of offensive weapons by the other side. There would be no need to react by deploying offensive weapons if the other side had an effective defence but little offensive capability. In fact, to do so would be counter-productive in an age when defence is more cost-effective than offence.

A non-provocative defence would greatly improve the atmosphere for arms control in Europe. An effective conventional defensive deterrent would, by removing perceptions of threats, increase trust and confidence between East and West and encourage détente. Arms control negotiations will only succeed if they take place in conditions of détente.

The adoption of a non-provocative defence by NATO European countries would increase stability in Europe during a crisis because the Soviets would not fear an attack and there would, therefore, be time to solve the crisis by political and diplomatic means. It would also reduce NATO's reliance on mobilization. Since premature mobilization is likely to encourage the other side to make a pre-emptive attack, to reduce the need for it would again increase crisis stability.

The risk of unintentional or accidental war would then be considerably reduced. It is precisely this risk that is increased significantly by the deployment of new nuclear weapons, like Pershing–IIs, in Europe.

Because the sort of non-provocative defence described here uses modern developments in defensive weaponry and its supporting technologies to best advantage, it would maximize NATO's advantage in microelectronic, microprocessor and missile guidance technologies. NATO is likely to maintain its lead for the foreseeable future.

A policy that provides effective self-defence without reliance on nuclear weapons but is not threatening is socially and morally acceptable. It is also unambiguously legal. The use of nuclear weapons, on the other hand, is illegal under a number of international treaties and conventions.

It can be concluded that non-provocative, non-nuclear defence satisfies all the six criteria listed above for an acceptable military posture for NATO countries. But it must be emphasized that it would not make war in Europe more acceptable. In fact, the choice for Europe is not between nuclear war and conventional war but between war and no war.

In this connection, the great destructive power of modern conventional weapons must be remembered. Some types of conventional weapons – like, for example, the fuel air explosive, in which an aerosol cloud of a very explosive mixture is ignited when the cloud is of optimum size – have the same explosive power as low-yield nuclear weapons. A full-scale conventional war in Europe would cause unimaginable damage, even if no nuclear weapons were used.

Any war in Europe is most likely to escalate to an all-out strategic nuclear war, even if there were no nuclear weapons deployed there. The probability of any war breaking out in Europe should, therefore, be minimized. And this is just what a non-provocative, non-nuclear defence would achieve.

Barriers to non-provocative defence

If a non-provocative defence is such a good idea, why has it not been adopted? There are, unfortunately, a number of barriers to its adoption.

One is the 'German problem'. West Germany opposes any change in NATO's policy of 'forward defence' which pays lip-service to the idea that, in a war in Europe, absolutely no West German territory should be given up. West German leaders believe that the adoption of a defence in depth would weaken forward defence and, therefore, they oppose it.

The West German attitude is not consistent. A severe weakness in NATO's current posture is that West Germany's border with the East is not fortified. There are no obstacles to an invading force in place, which, of course, from a military point of view, is nonsense. The reason for this extraordinary situation is that West German politicians still argue for unification with East Germany and claim that fortifying the border would imply a permanent frontier and make unification more difficult to achieve. They therefore object to a non-provocative defence policy that requires an effective barrier zone at the East–West German frontier.

Another obstacle to the adoption of non-provocative defence is that military tactics always lag behind technology. History shows that it takes the military some time to realize the significance of advances in military technology and then to make use of them.

For example, during the 1930s, military men, with few exceptions, fiercely resisted giving up cavalry in favour of tanks. So much so, that, difficult though it is to believe, Polish cavalry on horseback actually charged German tanks when Germany invaded Poland in 1939. Many military men now find it equally hard to face the fact that, for example, main battle tanks are obsolete. And, as one might expect, pilots are reluctant to give up manned aircraft, and their jobs, in favour of remotely piloted vehicles.

Another serious barrier is bureaucratic. Ministries of Defence budget for weapon procurements years in advance and it is difficult to change the minds of bureaucrats once they are made up. Bureaucratic inertia is hard to overcome.

For economic reasons, however, the weapons procured for the military in European NATO countries will be increasingly those favoured by the advocates of non-provocative defence. Already, the military cannot afford the latest main battle tanks, like the Challenger and Leopard–II, in strategically significant numbers. And few believe that another main battle tank will be developed.

Also, one can assume that no more multi-role combat aircraft, like the Tornado, will be produced. If NATO countries procure new combat aircraft they will probably be single-role interceptors or close-support ground attack aircraft. And most European navies have already recognized that large warships are obsolete and have got rid of them.

As the *weapons* change, it is to be hoped that the military will draw the logical conclusions and change their *total posture* to a conventional defensive deterrent. They may then recognize that nuclear weapons are unusable and no longer rely on them.

There is some, but not much, public resistance to non-provocative defence. This comes from pacifists, whose principles do not allow them to depend on any weapons even for defence, and from those against technology. The latter are put off by the fact that the concept utilizes some of the latest technologies, even though these are only defensive ones.

Conclusions

Weapons are, of course, not the only factor in a military posture. The tactics, operational skills and motivation of the troops are perhaps more important. And morale inevitably plays a critical role. The success of a military policy will be determined by whether or not the armed forces understand their function and are convinced of its usefulness.

Today's NATO policy is so incredible that the military forces cannot be expected to be motivated by it. Moreover, the threat to use nuclear weapons soon after a war begins in Europe involves preparing NATO troops for actions that civilized societies regard as morally unacceptable and illegal. On the other hand, conventional non-provocative defensive deterrence is totally consistent with the universally recognized right of self-defence and is, therefore, morally acceptable and unambiguously legal. And it is militarily credible. For these reasons, NATO populations in general and the armed forces in particular should welcome it.

Index

INTERNATIONAL RELATIONS IN A CHANGING WORLD

Joseph Frankel

In this 'comprehensive and compendious piece of work' *Times Literary Supplement,* Professor Frankel combines a description of the foreign policies of the major world powers with a fascinating analysis of the diplomatic setting in which such policies have to be formulated. He demonstrates the growing recognition of the need for some form of global or regional regimes to tackle the problem of *International Relations in a Changing World.*

Professor Frankel pays particular attention to the state powers of the USA, Soviet Union, and China, and considers the relevance of international institutions such as the League of Nations and the United Nations in the prospects for solving problems caused by the conflicts between states.

A Opus book

WAR IN EUROPEAN HISTORY

Michael Howard

This book offers a fascinating study of warfare as it has developed in Western Europe from the warring knights of the Dark Ages to the nuclear weapons of the present day, illustrating how war has changed society and how society in its turn has shaped the pattern of warfare.

'Wars have often determined the character of society. Society in exchange has determined the character of wars. This is the theme of Michael Howard's stimulating book. It is written with all his usual skill and in its small compass is perhaps the most original book he has written. Though he surveys a thousand years of history, he does so without sinking in a slough of facts and draws a broad outline of developments which will delight the general reader.' A. J. P. Taylor, *Observer*

'It is, at one and the same time, the plain man's guide to the subject, an essential introduction for serious students, and in its later stages a thought-provoking contribution.' Michael Mallet, *Sunday Times*